The

Reference

Shelf

Biotechnology

Edited by Lynn Messina

The Reference Shelf
Volume 72 • Number 4

The H.W. Wilson Company
New York • Dublin
2000

The Reference Shelf

The books in this series contain reprints of articles, excerpts from books, addresses on current issues, and studies of social trends in the United States and other countries. There are six separately bound numbers in each volume, all of which are usually published in the same calendar year. Numbers one through five are each devoted to a single subject, providing background information and discussion from various points of view and concluding with a subject index and comprehensive bibliography that lists books, pamphlets, and abstracts of additional articles on the subject. The final number of each volume is a collection of recent speeches, and it contains a cumulative speaker index. Books in the series may be purchased individually or on subscription.

Library of Congress has cataloged this serial title as follows:

Biotechnology/ edited by Lynn Messina.
    p. cm.— (The reference shelf ; v. 72, no. 4)
    Includes bibliographical references and index.
    ISBN 0-8242-0985-0 (pbk)
    1. Biotechnology—Popular works. 2. Genetic engineering—Popular works. I. Messina, Lynn.
II. Series.

TP248.215 .B56 2000
660.6—dc21   00-043622

Visit H.W. Wilson's Web site: www.hwwilson.com

Printed in the United States of America

Contents

Preface

Saying the words "clone," "genetic engineering," "xenotransplantation," and "cybernetics" to a group of people will elicit a variety of responses. In this age of mass media and easy access to information, even the least educated among us will be familiar with at least two of these four terms from the field of biotechnology. The more literary-minded might react with gleeful wonder, as images are conjured from Mary Shelley's *Frankenstein* or Aldous Huxley's *Brave New World*. A similar amusement might grip pop culture afficionados familiar with such TV classics as *Star Trek* and *The X-Files*. Those guided by deep religious faith, by contrast, might respond with fear, anger, or both, indignant at the thought that human beings should presume to "play God" by using their knowledge to interfere with the Divine laws of the natural world. Still others might express concern that, in the wrong hands, such technology could be used to tyrannize the human race. Despite the diversity of the public's response, no one can dispute that developments like those above in the field of biotechnology are changing the way human beings look at themselves and the world. What was once the stuff of our wildest dreams—the cloning of a human being, the building of a bionic man or woman, the curing of cancer, to name a few—are now within the realm of possibility. Despite the warnings of some that our Promethean endeavors to gain complete knowledge and control of human life will cause our downfall, biologists and engineers continue to make significant discoveries and produce new innovations at an astonishing rate. The articles included here attempt to explore many of those developments. Proceeding from inside the human body outward, this volume begins with a section on the latest research into the most elemental part of human beings—the human genome—and concludes with what most of us put into our bodies—genetically modified food, currently a subject of intense debate in the U.S. and around the world. The articles examine a variety of developments in biotechnology. Some are news items; others are more analytical and consider the ethical and moral issues raised by the discoveries.

The book's first section is devoted to the Human Genome Project, an extraordinary undertaking which, when completed, could supply the proverbial Holy Grail of human life. Through the mapping of human chromosomes, scientists hope not only to locate the genes that cause such conditions as Down syndrome and diseases like Alzheimer's and cancer, but also to use that knowledge to prevent or cure those and other afflictions. Robert Cooke's article provides a thorough overview of this enormous task, including a look at the race between the National Institutes of Health and the private biotech firm

Celera Genomics to complete the project first. Francis Collins of the NIH then answers some of the most relevant and commonly-asked questions about the project in an article written with Karin G. Jegalian for *Scientific America.* Then, Stanford University professor Robert Sapolsky, in his essay titled "It's Not 'All in the Genes,'" explores the nature/nurture debate and addresses the notion that our genes determine our behavior and our future lifestyle. Reports on the mapping of the worm and fruit fly genomes are also included due to the creatures' significance as the first complex organisms to be sequenced and because of what scientists anticipate they can learn about human life from the results. Then, writing for *Newsday,* Ellen Goodman probes the issue of selling and patenting sections of the human genome by individuals hoping to make a profit from the research.

Because each individual's DNA provides a unique blueprint of his or her genetic makeup, DNA tests are already being used both for identification and as a means of diagnosing and predicting a person's present or future medical condition. The second section therefore addresses the use of such tests by law enforcement officials to convict criminals, a process discussed in Ronald Bailey's article for *Reason* and by prospective parents to determine whether or not to adopt children with potential health problems, a controversial practice discussed by *Boston Globe* writer Richard Saltus. In a related issue, Kristi Coale, writing for *Salon,* questions the wisdom of those who decide whether or not to have children based on the genetic composition of either themselves, the fetuses, or both.

Many doctors are already using knowledge from genetic tests to treat and, in many cases, cure individuals of some of the most life-threatening illnesses, and these treatments are at the center of this book's third section. *Science* magazine's Gretchen Vogel discusses what that publication considers the most important breakthrough of 1999—stem cell research. The political ramifications of research into stem cells and other innovations that may eliminate illness altogether are then discussed in another piece by Ronald Bailey for *Reason,* an article complemented by the *Wall Street Journal*'s Sherwin B. Nuland, who wonders if we really *want* to live longer, albeit healthier lives. The section ends on a hopeful note, with the story of 13-year-old Ashanthi DeSilva, whose life has been saved by the controversial treatment known as gene therapy.

With even the most incredible and wondrous innovation often comes fear, however, and no development in the field of biotechnology has inspired greater fear and controversy than cloning, the subject of the volume's fourth section. From *New York Times* science writer Gina Kolata's 1997 report on the ethical implications of the creation of Dolly the sheep, we move to a recent *Los Angeles Times* interview with Dolly's "father," the Scottish scientist Ian Wilmut, who continues to see great hope and possibility in cloning. One of the most commonly discussed of those possibilities is the use of human clones for organ transplants, which is explored by Charles Krauthammer in *Time*. Opposition to human cloning continues, however, as we see in the response from Stephen G. Post, whose article in *America* offers a thorough and measured consideration of the moral and spiritual ramifications of cloning from a Judeo-Christian perspective.

Until human cloning becomes an accepted means of generating spare parts for those in need of organ or limb transplants, scientists continue to devise a means of producing those parts in the laboratory, and this is the subject of the book's fifth section. Nearly 25 years after the 1970s TV show *The Six Million Dollar Man* and its spinoff *The Bionic Woman* first aired, scientists have constructed a variety of replacement parts to help rebuild injured human bodies. Perhaps the most coveted such creation would be an implantable heart made of human cardiac cells, as discussed in an article for the *Chronicle of Higher Education*. An article from the *New York Times* on the "whole body transplant" describes what may be the strangest of the current medical innovations, while the development of brain-controlled prostheses like those reported in *Science News* offer much more practical solutions for a number of paraplegics. Also included here is an examination of the promise and perils of xenotransplantation, by which animal organs are implanted into human beings, a process necessitated by the shortage of human organ donations. Kevin Warwick's account of his experience living with a silicon chip implant completes the section, as Warwick looks at his life as a cyborg—a human being whose thoughts and emotions are linked remotely to a computer.

The book's sixth and final section considers the debate over the marketing and consumption of genetically modified organisms, or GMOs, beginning with Robert Paarlberg's review of both sides of the issue in *Foreign Affairs*. James Freeman in *USA Today* then argues that fears of so-called "Frankenfoods" posing a dangerous health risk are largely unfounded. A final article from the *New York Times* discusses attempts by farmers to devise new ways of genetically altering crops so as to reduce the perceived risk to those who consume the foods that are made from them.

I wish to thank the authors and publications that granted permission for their work to be reprinted in this volume. I would also like to acknowledge Dr. Lauren Albert, Dr. Signe Kelker, Dr. Eleanor O'Rangers, Denise Bonilla, Sara Yoo, and Sandra Watson, all of whom provided invaluable assistance in the compilation of this book.

Lynn Messina
July 2000

I.

The Human Genome Project

I

The Human Genome Project

Editor's Introduction

On the morning of June 26, 2000, those tuning into their favorite news program were greeted with a report of astounding significance for humankind: A rough draft of the human genome had been completed. As journalists and news readers compared the achievement to the first moon landing and the discovery of DNA's structure almost 50 years earlier, President Clinton spoke in even loftier terms, proclaiming, "Today we are learning the language in which God created life." It is difficult to overstate the importance of this announcement, which was made jointly by the governments of the United States and Great Britain. The possibilities are both thrilling and frightening. With the knowledge gained from studying the genome, scientists can understand not only how the human body is put together, but also how to repair physiological imperfections that may prevent a person from living a full, healthy life. On a more worrisome note, scientists may one day be able to locate the precise genes for our most valuable and enviable talents, which could pave the way for the engineering of children or, more ominously, discrimination against those who may be considered genetically inferior. Yet, as the articles in this first section demonstrate, from the beginning, the goals of the Human Genome Project have been more altruistic than sinister, as researchers in this new field of biotechnology strive to understand our genetic composition in order to cure disease and improve the quality of human life.

The first article in this section, Robert Cooke's story for *Newsday* entitled "Milestone for Humanity," covers that historic June morning when it was announced that researchers had a completed map of 90% of the human genome. Cooke describes this 15-year project to achieve what he calls "the pinnacle of human self-knowledge," including the fierce competition and somewhat surprising eleventh-hour collaboration between the National Human Genome Research Institute, a publicly funded consortium headed by Dr. Francis S. Collins, and the private firm Celera Genomics, led by Dr. J. Craig Venter. As he reviews the drama behind the scenes, Cooke also offers a lucid, concise explanation of the complex medical and scientific concepts involved in this scientific "milestone," along with its significance for humankind.

The National Institutes of Health's Dr. Collins then offers his own perspective on this undertaking in an essay for *Scientific American* that he co-authored with Karin G. Jegalian, one of the magazine's writers. "Deciphering the Code of Life," which benefits from Collins's unique position on the research team, asks and then answers several intriguing questions about this new field called genomics, among them, "Will synthetic life-forms be produced?"; "Will understanding the human genome transform preventive, diagnostic and therapeutic medicine?"; and "Will we reconstruct accurately the history of human populations?" In this article, the writers effectively demonstrate that the study of genomics generates a "virtuous cycle" in which "the more we learn, the more we will be able to extrapolate, hypothesize and understand."

One of the public's biggest fears concerning the knowledge obtained from the study of the human genome is that we will discover that our character, abilities, and potential achievements—indeed, what we are fundamentally—are ultimately determined by our genes. In his essay for *Newsweek*, "It's Not 'All in the Genes,'" Robert Sapolsky, a professor of biological sciences and neurology at Stanford University, seeks to allay those fears by explaining how "nurture reinforces or retards nature." Given the manner in which our genes interact with our bodies' cells, he argues, and, in turn, how those cells interact with our emotions and environment, our genes cannot be considered completely responsible for who and what we are, nor what we will be.

Before the completion of the Human Genome Project, two other complex life forms had been sequenced, the worm called *Caenorhabiditis elegans*—a.k.a. *C. elegans*—and the fruit fly. As J. Travis explains in *Science News*, from the genome for *C. elegans*, which was the first animal to be sequenced, scientists can learn a great deal about human development; its transparent body enables researchers to study the connections between genes and processes like embryonic development and the nervous system. In *Science*, Elizabeth Pennisi reports on the sequencing of the fruit fly genome, which is significant primarily because of the method by which it was completed. The "shotgun" approach employed by the team, which included Celera Genomics's J. Craig Venter, had been deemed "unworkable," but its success in this project now suggests to researchers another tool that may be available to sequence the human genome.

Now that this invaluable knowledge about our genetic composition has been uncovered by scientists, the next step is to make it available to researchers, a topic addressed by Ellen Goodman of the *Boston Globe*. In her article entitled "Our Genes for Sale—Get 'em While They're Hot," she reflects on the practice of selling patents on human genes to private corporations as a means of stimulating biomedical and genetic research. In this way, private corporations and individuals may own the rights to any discoveries made from a particular section of the human genome, along with the profits earned from drugs and other forms of medical treatment developed from that section. After addressing some of the philosophical questions involved in this approach, Goodman presents Dr. Collins's view on the issue, including his use of "a toll booth analogy" to explain its necessity and the ultimate benefits for humanity.

"Milestone for Humanity"[1]

Scientists Decode Blueprint of the Human Body

BY ROBERT COOKE
NEWSDAY, JUNE 27, 2000

The genome is in. Except for a few jots, tittles and i's yet to be dotted, all of the basic links in the chain of DNA that spells out humanity's genetic endowment are now known, ready to be studied in exquisite detail, scientists announced yesterday.

The achievement—a true milestone in science—was announced at a White House news conference by President Bill Clinton, British Prime Minister Tony Blair—via satellite hookup—and scientists who led the international gene-finding venture.

Before an audience crowded with leading gene researchers, including Nobel laureate James Watson of Cold Spring Harbor Laboratory, Clinton stated:

"Today we are learning the language in which God created life. We are gaining ever more awe for the complexity, the beauty, the wonder of God's most divine and sacred gift.

"With this profound new knowledge, humankind is on the verge of gaining immense new power to heal. Genome science will have a real impact on all our lives, and even more on the lives of our children. It will revolutionize the diagnosis, prevention and treatment of most, if not all, human diseases."

Blair, speaking from 10 Downing St. in London, added: "Let us be in no doubt about what we are witnessing today; a revolution in medical science whose implications surpass even the discovery of antibiotics."

The announcement marks a benchmark in genome research. Nothing is actually completed, and an enormous amount of work remains. But the public project had said that the achievement of 90 percent of the mapping of our genetic content would represent a milestone.

And yesterday they celebrated, telling the world they had achieved their goal—and then some.

As genome data pour in, researchers expect to gain a better foothold in cancer treatment—they talk even of reducing the fatality rate to zero someday—as well as treating some inherited diseases

---

1. Copyright © 2000 Newsday, Inc. Reprinted with permission.

that have forever been unapproachable. Genetic findings may also make it clear why some people are more susceptible to infections than others.

Flanking Clinton during the announcement were two scientists who had been cast as bitter rivals: J. Craig Venter, president of the private company Celera Genomics Corp., and Francis Collins, director of the federally funded Human Genome Project. After months of conflict and contention, the two had finally put down their cudgels and agreed to cooperate, if not collaborate.

Venter said that Celera's version of the human genome map is 99 percent complete, with numerous small gaps still existing in parts of chromosomes that have been hard to sequence. Collins said the Human Genome Project has 97 percent of the genome sliced up, reassembled and mapped, with about 85 percent sequenced in fine detail. Sequencing involves reading the genome, noting the identity and location of each chemical "letter" in the genetic codebook.

Both teams—the private and the public—plan to publish their results simultaneously later this year, and then perhaps cooperate in working out how their independent maps compare. The Human Genome Project now estimates there are 3.15 billion sub-units or letters of DNA, called base-pairs, in the genome. Venter's team estimates almost the same, 3.12 billion base-pairs.

What is still not known, however, is how many human genes there really are. Collins said yesterday the new data indicate there are at least 38,000 human genes, although there may actually be far more, perhaps 100,000.

Both gene-seeking teams also promised that by the end of the year all of their data will be available to scientists worldwide on the Internet. That is a departure from Celera's original position, which was that its data would be withheld for a substantial time while its scientists searched through it for genes with commercial value. Collins' team has made a point of making its data public immediately, dumping it into the genome database daily for use by scientists everywhere.

In the nine months that Celera has been seeking genes with commercial value, Venter said, the company has filed for patents on about two dozen individual genes. But he said the company's main goal—the way it expects to make money—is to become an information resource, a place where the genomes of humans, mice, cats, dogs, rats and many other organisms can be compared. Permission to use that resource is being sold via subscription to major pharmaceutical companies, biotechnology firms and some universities.

The result, he added, should be that research that once took weeks, months or years to perform in the laboratory can now be done immediately in the computer, perhaps in just seconds. For research companies, the savings would be enormous.

Both Venter and Collins also tried to overcome the impression that they have been warring competitors racing to be first to "capture" the human genome. "The focus on a race and the personality issues . . . has in many ways been a disservice," Collins said. "I don't think the animosity or hostility was anything approaching the way it was described in some pieces that Craig and I have had to read."

In any case, Collins added, "This, after all, is a noble enterprise. Sequencing our genome is not something that is tarnished in some way by what appears to be a cat fight among the people who are involved."

In fact, Collins said, "It is humbling for me and awe-inspiring to realize that we have caught the first glimpse of our own instruction book, previously known only to God. What a profound responsibility it is to do this work. Historians will consider this a turning point."

All parties—Clinton, Blair, Venter and Collins—emphasized over and over again that effort must now go into making sure the new genome data are used responsibly. There is much fear that insurance companies and employers will use such information as a means for screening high-risk people out of jobs or eligibility for health coverage.

Such anxiety about the human genome is well-placed. As scientists scramble to unravel every last bit of information in the human genes, what they're digging into is, in fact, the common denominator of everyone.

In effect, our genome is a fundamentally important document written in an obscure language, telling of human life's history on Earth, where we came from and when, and riddled with errors that sometimes make us sick.

Each human is born with the same basic set of genetic information, but with small variations that make us individuals. And these tiny differences will soon be quickly readable, probably on electronic chips, to assess our inborn strengths and weaknesses.

Mapping the entire human genome, Collins said, "will lead scientists to previously unimagined insights, and from there to the common good, a new understanding of the genetic contributions to human disease."

Also, Collins said, its size, complexity and ambition make the Human Genome Project "an audacious program" that is well ahead of its expected timetable, and coming in below the esti-

> *"It is humbling for me and awe-inspiring to realize that we have caught the first glimpse of our own instruction book, previously known only to God."* —**Francis Collins, director of the Human Genome Project**

mated cost of $3 billion. He said yesterday that the cost so far has been only about $300 million.

Collins recently urged hundreds of researchers at a major genome meeting to help people understand what is going on, and where genome research is leading. He said his fellow scientists need to open up and communicate publicly about their work in the pursuit of genes. Speaking at the Cold Spring Harbor Laboratory in May, he said scientists need to emerge from their laboratories and share their genetic wisdom, to be ambassadors for their science.

Helping people understand the genome project will be a big job, however, because the genome itself is in many ways still a major mystery. For example, if printed out word-for-word, the human genome would fill the equivalent of 200 Manhattan telephone books, each 100 pages thick. It would also take about 19 years if one tried to read out every bit of chemical information stored therein.

Helping people understand the genome project will be a big job, however, because the genome itself is in many ways still a major mystery.

But that's exactly what scientists are now doing—reading this encyclopedia of life—in a big hurry. According to geneticist/mathematician Eric Lander, members of the publicly funded genome consortium are "sequencing" the human genome with extraordinary speed. By this spring, he said, 10,000 bits of information, base-pairs, were being read into the computer databank, GenBank, every minute.

Base-pairs are the individual links, akin to letters, that are strung together to "spell" a gene's biochemical message. The arrangement of these base-pairs governs how amino acids, the building blocks of proteins, get linked together to create the proteins in skin, bone, blood and other bodily materials.

One question that is often raised concerns whose genome is it, anyway? According to John Daley, a biologist at the Whitehead Institute in Cambridge, Mass., much of the DNA used in the public project "came from a man and a woman living in Buffalo, N.Y." They have not been identified, but they were chosen decades ago when the first gene-splicing experiments were being done, and their DNA was put into a commonly used gene "library." Other genetic material used in the project came from people in the Los Angeles area.

The researchers at Celera Genomics Corp., however, are using different sets of DNA they collected themselves, from unidentified people; three women, two men, and including people of Hispanic, Asian, caucasian and African-American background.

"We did that to help illustrate the concept of race has no genetic or scientific basis," Venter explained. In the Celera genomes, "there's no way to tell one ethnicity from another."

Even as the genome is being deciphered, scientists worldwide have not waited to begin digging into the accumulating data, using it to further their own studies of genetic diseases, metabolic processes and even the history of the human race.

At the same time, several private companies, including Celera, of Rockville, Md., were sequencing on their own, even promising to get the job done before the publicly funded Human Genome Project could finish. And despite scientists' claims that it's not really a race, they've acted like it is.

"We are in a hurry," biologist Robert Waterston said earlier. "We want to get it done and get it out there" where the world's scientists can use it. "We're impatient people anyway."

Certainly the stock market has noticed, even if scientists argue it wasn't actually a race. The shares of Celera's parent company, Perkin-Elmer Inc., have yo-yoed up and down dramatically each time there's been a hint that one side or the other is ahead. Investors are speculating that Celera can find and patent enough of the genome to become very, very rich on genome information, and on gene-based biomedical products. Yesterday shares of Celera rose as high as $135 before closing at $112.

According to Waterston, the genome's real completion date—when virtually all of the blanks are filled in and the sequence is 99.99 percent done—will come in the year 2003. By then, it may be known exactly where each gene is located within these 3 billion base-pairs, and scientists will be learning how the genes work together under precise control.

Richard Gibbs, director of the genome sequencing center at Baylor College of Medicine in Houston, said that once the entire genome is in place, "it's only the beginning, not the end. At some point the baton will be passed from the whole genome search to the skilled individual researchers who are interested in deep understanding" of a given disease, or a physical trait.

"The physician-scientists who are studying heart disease," for instance, Gibbs said, "will know as far as it's possible to know the fine differences between the various clinical pathologies" in heart ailments. "Some are reaching into the raw data to make those distinctions already."

Baylor is one of five major labs in the public project. The others are the Sanger Center in Cambridge, England, the Whitehead Institute in Cambridge, Mass., Washington University in St. Louis, and the Department of Energy sequencing laboratory in Walnut Creek, Calif.

Almost by accident, the timing of completion of the human genome will be appropriate. When the complete, well-founded genome is known in 2003, it will also be the 50th anniversary of

A GENOME PRIMER

Amino Acid

The molecular building block of proteins. The proteins in our bodies are composed of various combinations of 20 different amino acids.

Base pair (bp)

Two DNA subunits held together by weak chemical bonds between their nitrogen-containing components. In nature, the component, or base, called guanine pairs with cytosine (GC or CG), and the base adenine pairs with thymine (AT or TA). The bonds between multiple base pairs hold two strands of DNA together in the shape of a double helix.

Chromosome

A structure in the nucleus of a cell that contains the cell's genetic information. We have 23 pairs of chromosomes, each composed of DNA subunits whose sequence determines our entire array of genes.

Cloning

Producing multiple, exact copies of a single gene, other DNA segments, an entire cell or a complete organism. Cloned collections of DNA are called clone libraries and are useful tools in helping scientists piece together our genetic information.

DNA

An abbreviation for deoxyribonucleic acid. The long, double-stranded molecule coiled inside each cell consists of individual subunits called nucleotides that "spell out" our genetic information (See Nucleotide).

Gene

An ordered string of DNA nucleotides that we inherit from our parents. Each gene has a unique location on a particular chromosome and acts as the blueprint for producing a specific protein.

Gene Mapping

Determining the location of a particular DNA sequence on one of the 23 pairs of chromosomes that make up our genetic material.

Gene Therapy

Transferring new DNA or an entire gene into an individual, usually with the goal of replacing that individual's damaged or missing gene and thus partially or completely restoring the intended function.

Genetic Code

The sequence of DNA nucleotides, read as three-letter words called codons, that acts as a template for ordering amino acids to form proteins.

Genome

The entire complement of our genetic material. The human genome consists of about 3 billion base pairs dispersed among Chromosomes X, Y, and 1 through 22.

Nucleotides

DNA or RNA subunits containing a phosphate molecule, a sugar molecule and one of four nitrogen-containing molecules called bases. Each DNA nucleotide contains either an adenine, thymine, guanine or cytosine base, often referred to by the single letters A, T, G or C.

Nucleus

The cellular compartment that contains our genetic information.

PCR

Abbreviation for Polymerase Chain Reaction. Scientists use this technique to rapidly increase the amount of a specific DNA sequence or to detect the existence of a defined sequence within a particular DNA sample.

Recombinant DNA

DNA molecules with different origins that have been joined to form a new chimeric chain.

RNA

An abbreviation for ribonucleic acid. Structurally similar to DNA, this molecule plays an important role in producing proteins from DNA blueprints.

Sequencing

Determining the relative order of nucleotides in a DNA or RNA molecule or the order of amino acids in a protein.

the discovery of DNA's molecular structure by Nobel laureates James Watson, Francis Crick and Maurice Wilkins, a landmark event in the history of science. And because of Watson's central role in uncovering the structure of DNA, and his high visibility as a scientist, he was chosen as the first director of the Human Genome Project.

According to Collins, director of the Human Genome Project for the National Institutes of Health, the genome data collected so far haven't solved a fundamental riddle: How many genes exist, anyway? Estimates have ranged between 35,000 and 140,000 human genes. The figure is so imprecise because recent mapping of chromosome 21 found far fewer genes than expected, only 225 instead of about 500.

In contrast, the sequencers at Celera recently suggested there may be more genes than expected hidden on the chromosomes, as many as 140,000. So in jest, Collins has been presiding over a lottery that will pay off in 2003, when the genome is done.

At the annual genome conference at Cold Spring Harbor in May, Collins began asking his colleagues to submit guesses at $1 each now, and $5 each later into a pool that will be opened at the annual genome meeting in 2003. The winner will be the scientist(s) who comes closest to guessing the actual number of human genes.

This lack of firm knowledge about how many human genes actually exist "is pretty striking," Waterston said. The puzzle exists because "it's not a simple thing to find genes. Only about 3 percent of the genome is used in genes" that actually specify the construction of proteins. The rest includes a huge amount of DNA of unknown function that sits between genes and within genes. And there are many known genes that seem to control the activities of others.

According to Daley, at the Whitehead Institute, the pursuit of a single gene is an arduous task. He compared the search to looking at the Earth from outer space, focusing first on Lake Michigan, then finding Chicago on the shore, narrowing the view down to one neighborhood, finding a specific house, spotting a person sitting in the back yard, and then examining the hairs on the back of her hand.

Deciphering the Code of Life[2]

By Francis S. Collins and Karin G. Jegalian
Scientific American, December 1999

When historians look back at this turning of the millennium, they will note that the major scientific breakthrough of the era was the characterization in ultimate detail of the genetic instructions that shape a human being. The Human Genome Project— which aims to map every gene and spell out letter by letter the literal thread of life, DNA—will affect just about every branch of biology. The complete DNA sequencing of more and more organisms, including humans, will answer many important questions, such as how organisms evolved, whether synthetic life will ever be possible and how to treat a wide range of medical disorders.

The Human Genome Project is generating an amount of data unprecedented in biology. A simple list of the units of DNA, called bases, that make up the human genome would fill 200 telephone books—even without annotations describing what those DNA sequences do. A working draft of 90 percent of the total human DNA sequence should be in hand by the spring of 2000, and the full sequence is expected in 2003. But that will be merely a skeleton that will require many layers of annotation to give it meaning. The payoff from the reference work will come from understanding the proteins encoded by the genes.

Proteins not only make up the structural bulk of the human body but also include the enzymes that carry out the biochemical reactions of life. They are composed of units called amino adds linked together in a long string; each string folds in a way that determines the function of a protein. The order of the amino acids is set by the DNA base sequence of the gene that encodes a given protein, through intermediaries called RNA; genes that actively make RNA are said to be "expressed."

The Human Genome Project seeks not just to elucidate all the proteins produced within a human but also to comprehend how the genes that encode the proteins are expressed, how the DNA sequences of those genes stack up against comparable genes of other species, how genes vary within our species and how DNA sequences translate into observable characteristics. Layers of information built on top of the DNA sequence will reveal the

---

knowledge embedded in the DNA. These data will fuel advances in biology for at least the next century. In a virtuous cycle, the more we learn, the more we will be able to extrapolate, hypothesize and understand.

By 2050 we believe that genomics will be able to answer the following major questions:

Will the three-dimensional structures of proteins be predictable from their amino acid sequences?

The six billion bases of the human genome are thought to encode approximately 100,000 proteins. Although the sequence of amino acids in a protein can be translated in a simple step from the DNA sequence of a gene, we cannot currently elucidate the shape of a protein on purely theoretical grounds, and determining structures experimentally can be quite laborious. Still, a protein's structure is conserved—or maintained fairly constantly throughout evolution—much more than its amino acid sequence is. Many different amino acid sequences can lead to proteins of similar shapes, so we can infer the structures of various proteins by studying a representative subset of proteins in detail.

Recently an international group of structural biologists have begun a Protein Structure Initiative to coordinate their work. Structural biologists "solve" the shapes of proteins either by making very pure crystals of a given protein and then bombarding the crystals with x-rays or by subjecting a particular protein to nuclear magnetic resonance (NMR) analysis. Both techniques are time-consuming and expensive. The consortium intends to get the most information out of each new structure by using existing knowledge about related structures to group proteins into families that are most likely to share the same architectural features. Then the members of the consortium plan to target representatives of each family for examination by painstaking physical techniques.

As the catalogue of solved structures swells and scientists develop more refined schemes for grouping structures into a compendium of basic shapes, biochemists will increasingly be able to use computers to model the structures of newly discovered—or even wholly invented-proteins. Structural biologists project that a total of about 1,000 basic protein-folding motifs exist; current models suggest that solving just 3,000 to 5,000 selected structures, beyond the ones already known, could allow researchers to deduce the structures of new proteins routinely. With structural biologists solving more than 1,000 protein structures every year and with their progress accelerating, they should be able to complete the inventory not long after the human genome itself is sequenced.

Will synthetic life-forms be produced?

Whereas structural biologists work to group proteins into categories for the practical aim of solving structures efficiently, the fact that proteins are so amenable to classification reverberates with biological meaning. It reflects how life on the earth evolved and opens the door to questions central to understanding the phenomenon of life itself. Is there a set of proteins common to all organisms? What are the biochemical processes required for life?

Already, with several fully sequenced genomes available— mostly from bacteria—scientists have started to take inventories of genes conserved among these organisms, guided by the grand question of what constitutes life, at least at the level of a single cell.

If, within a few years, investigators can expect to amass a tidy directory of the gene products—RNA as well as proteins—required for life, they may well be able to make a new organism from scratch by stringing DNA bases together into an invented genome coding for invented products. If this invented genome crafts a cell around itself and the cell reproduces reliably, the exercise would prove that we had deciphered the basic mechanisms of life. Such an experiment would also raise safety, ethical and theological issues that cannot be neglected.

Will we be able to build a computer model of a cell that contains all the components, identifies all the biochemical interactions and makes accurate predictions about the consequences of any stimulus given to that cell?

In the past 50 years, a single gene or a single protein often dominated a biologist's research. In the next 50 years, researchers will shift to studying integrated functions among many genes, the web of interactions among gene pathways and how outside influences affect the system.

Of course, biologists have long endeavored to describe how components of a cell interact: how molecules called transcription factors bind to specific scraps of DNA to control gene expression, for example, or how insulin binds to its receptor on the surface of a muscle cell and triggers a cascade of reactions in the cell that ultimately boosts the number of glucose transporters in the cell membrane. But the genome project will spark similar analyses for thousands of genes and cell components at a time. Within the next half-century, with all genes identified and all possible cellular interactions and reactions charted, pharmacologists developing a drug or toxicologists trying to predict whether a substance is poisonous may well turn to computer models of cells to answer their questions.

Will the details of how genes determine mammalian development become clear?

Being able to model a single cell will be impressive, but to understand fully the life-forms we are most familiar with, we will plainly have to consider additional levels of complexity. We will have to examine how genes and their products behave in place and time—that is, in different parts of the body and in a body that changes over a life span. Developmental biologists have started to monitor how pools of gene products vary as tissues develop, in an attempt to find products that define stages of development. Now scientists are devising so-called expression arrays that survey thousands of gene products at a time, charting which ones turn on or off and which ones fluctuate in intensity of expression. Techniques such as these highlight many good candidates for genes that direct development and establish the animal body plan.

As in the past, model organisms—like the fruit fly *Drosophila*, the nematode *Caenorhabditis elegans* and the mouse—will remain the central workhorses in developmental biology. With the genome sequence of *C. elegans* finished, *Drosophila's* near completion, the full human sequence on the way by 2003 and the mouse's likely within four to five years, sequence comparisons will become more commonplace and thorough and will give biologists many clues about where to look for the driving forces that fashion a whole animal. Many more complete genomes representing diverse branches of the evolutionary tree will be derived as the cost of sequencing decreases.

Within 50 years, we expect comprehensive genomics-based health care to be the norm in the U.S.

So far developmental biologists have striven to find signals that are universally important in establishing an animal's body plan, the arrangement of its limbs and organs. In time, they will also describe the variations—in gene sequence and perhaps in gene regulation—that generate the striking diversity of forms among different species. By comparing species, we will learn how genetic circuits have been modified to carry out distinct programs so that almost equivalent networks of genes fashion, for example, small furry legs in mice and arms with opposable digits in humans.

Will understanding the human genome transform preventive, diagnostic and therapeutic medicine?

Molecular biology has long held out the promise of transforming medicine from a matter of serendipity to a rational pursuit grounded in a fundamental understanding of the mechanisms of life. Its findings have begun to infiltrate the practice of medicine; genomics will hasten the advance. Within 50 years, we expect comprehensive genomics-based health care to be the norm in the U.S.

We will understand the molecular foundation of diseases, be able to prevent them in many cases, and design accurate, individual therapies for illnesses.

In the next decade, genetic tests will routinely predict individual susceptibility to disease. One intention of the Human Genome Project is to identify common genetic variations. Once a list of variants is compiled, epidemiological studies will tease out how particular variations correlate with risk for disease. When the genome is completely open to us, such studies will reveal the roles of genes that contribute weakly to diseases on their own but that also interact with other genes and environmental influences such as diet, infection and prenatal exposure to affect health.

By 2010 to 2020 gene therapy should also become a common treatment, at least for a small set of conditions.

Within 20 years, novel drugs will be available that derive from a detailed molecular understanding of common illnesses such as diabetes and high blood pressure. The drugs will target molecules logically and therefore be potent without significant side effects. Drugs such as those for cancer will routinely be matched to a patient's likely response, as predicted by molecular fingerprinting. Diagnoses of many conditions will be much more thorough and specific than they are now. For example, a patient who learns that he has high cholesterol will also know which genes are responsible, what effect the high cholesterol is likely to have, and what diet and pharmacological measures will work best for him.

By 2050 many potential diseases will be cured at the molecular level before they arise, although large inequities worldwide in access to these advances will continue to stir tensions. When people become sick, gene therapies and drug therapies will home in on individual genes, as they exist in individual people, making for precise, customized treatment. The average life span will reach 90 to 95 years, and a detailed understanding of human aging genes will spur efforts to expand the maximum length of human life.

Will we reconstruct accurately the history of human populations?

Despite what may seem like great diversity in our species, studies from the past decade show that the human species is more homogeneous than many others; as a group, we display less variation than chimps do. Among humans, the same genetic variations tend to be found across all population groups, and only a small fraction of the total variation (between 10 and 15 percent) can be related to differences between groups. This has led some population biologists to the conclusion that not so long ago the human species was composed of a small group, perhaps 10,000 individuals, and that human populations dispersed over the earth only recently. Most genetic variation predated that time.

Clinton/Gore Administration U.S. Human Genome Project Funding History (Figures in Millions of $$) *The White House, March 14, 2000*			
Year	National Institutes of Health	Department of Energy	Total
1993	106.1	63.0	169.1
1994	127.0	63.3	190.3
1995	153.8	68.7	222.5
1996	169.3	73.9	243.2
1997	188.9	77.9	266.8
1998	218.3	85.5	303.8
1999	283.6	89.8	373.4
2000	335.9	88.9	424.8
2001 (Request)	357.7	90.3	448.0
			$2,641.9

Armed with techniques for analyzing DNA, population geneticists have for the past 20 years been able to address anthropological questions with unprecedented clarity. Demographic events such as migrations, population bottlenecks and expansions alter gene frequencies, leaving a detailed and comprehensive record of events in human history. Genetic data have bolstered the view that modem humans originated relatively recently, perhaps 100,000 to 200,000 years ago, in Africa and dispersed gradually into the rest of the world. Anthropologists have used DNA data to test cultural traditions about the origins of groups such as Gypsies and Jews, to track the migration of humans into the South Pacific islands and the Americas, and to glean insights into the spread of populations in Europe, among other examples. As DNA sequence data become increasingly easy to accumulate, relationships among groups of people will become clearer, revealing histories of intermingling as well as periods of separation and migration. Race and ethnicity will prove to be largely social and cultural ideas; sharp, scientifically based boundaries between groups will be found to be nonexistent.

By 2050, then, we will know much more than we do now about human populations, but a question remains: How much can be known? Human beings have mated with enough abandon that probably no one family tree will be the unique solution accounting for all human history. In fact, the history of human populations will emerge not as a tree but as a trellis where lineages often meet and mingle after intervals of separation. Still, in 50 years, we will know how much ambiguity remains in our reconstructed history.

Will we be able to reconstruct the major steps in the evolution of life on the earth?

Molecular sequences have been indispensable tools for drawing taxonomies since the 1960s. To a large extent, DNA sequence data have already exposed the record of 3.5 billion years of evolution, sorting living things into three domains— Archaea (single-celled organisms of ancient origin), Bacteria and Eukarya (organisms whose cells have a nucleus)—and revealing the branching patterns of hundreds of kingdoms and divisions. One aspect of inheritance has complicated the hope of assigning all living things to branches in a single tree of life. In many cases, different genes suggest different family histories for the same organisms; this reflects the fact that DNA isn't always inherited in the straightforward way, parent to offspring, with a more or less predictable rate of mutation marking the passage of time. Genes sometimes hop across large evolutionary gaps. Examples of this are mitochondria and chloroplasts, the energy-producing organelles of animals and plants, both of which contain their own genetic material and descended from bacteria that were evidently swallowed whole by eukaryotic cells.

Armed with techniques for analyzing DNA, population geneticists have for the past 20 years been able to address anthropological questions with unprecedented clarity.

This kind of "lateral gene transfer" appears to have been common enough in the history of life, so that comparing genes among species will not yield a single, universal family tree. As with human lineages, a more apt analogy for the history of life will be a net or a trellis, where separate lines diverge and join again, rather than a tree where branches never merge.

In 50 years, we will fill in many details about the history of life, although we might not fully understand how the first self-replicating organism came about. We will learn when and how, for instance, various lineages invented, adopted or adapted genes to acquire new sets of biochemical reactions or different body plans. The gene-based perspective of life will have taken hold so deeply among scientists that the basic unit they consider will very likely no longer be an organism or a species but a gene. They will chart which genes have traveled together for how long in which genomes. Scientists will also address the question that has dogged people since Charles Darwin's day: What makes us human? What distinguishes us as a species?

Undoubtedly, many other questions will arise over the next 50 years as well. As in any fertile scientific field, the data will fuel new hypotheses. Paradoxically, as it grows in importance, genomics itself may not even be a common concept in 50 years, as it radiates into many other fields and ultimately becomes absorbed as part of the infrastructure of all biomedicine.

How will individuals, families and society respond to this explosion in knowledge about our genetic heritage?

The social question, unlike the preceding scientific, technological and medical ones, does not come down to a yes-or-no answer. Genetic information and technology will afford great opportunities to improve health and to alleviate suffering. But any powerful technology comes with risks, and the more powerful the technology, the greater the risks. In the case of genetics, people of ill will today use genetic arguments to try to justify bigoted views about different racial and ethnic groups. As technology to analyze DNA has become increasingly widespread, insurers and employers have used the information to deny workers access to health care and jobs. How we will come to terms with the explosion of genetic information remains an open question.

Finally, will antitechnology movements be quieted by all the revelations of genetic science? Although we have enumerated so many questions to which we argue the answer will be yes, this is one where the answer will probably be no. The tension between scientific advances and the desire to return to a simple and more "natural" lifestyle will probably intensify as genomics seeps into more and more of our daily lives. The challenge will be to maintain a healthy balance and to shoulder collectively the responsibility for ensuring that the advances arising from genomics are not put to ill use.

It's Not "All in the Genes"[3]

The environment you grow up in is as important as your DNA in determining the person you ultimately become.

By ROBERT SAPOLSKY
NEWSWEEK, APRIL 10, 2000

It is no surprise that virtually every list that appeared of the most influential people of the 20th century included James Watson and Francis Crick, right up there alongside Churchill, Gandhi and Einstein. In discerning the double-helical nature of DNA, Watson and Crick paved the way for understanding the molecular biology of the gene, the dominant scientific accomplishment of the postwar era. Sequencing the human genome will represent a closure of sorts for the revolution wrought by those two geniuses.

At the same time, it's also not surprising that many people get nervous at the prospects of that scientific milestone. It will no doubt be a revolution, but there are some scary "Brave New World" overtones that raise fundamental questions about how we will think about ourselves. Will it mean that our behaviors, thoughts and emotions are merely the sum of our genes, and scientists can use a genetic road map to calculate just what that sum is? Who are we then, and what will happen to our cherished senses of individuality and free will? Will knowing our genetic code mean we will know our irrevocable fates?

I don't share that fear, and let me explain why. At the crux of the anxiety is the notion of the Primacy of Genes. This is the idea that if you want to explain some big, complex problem in biology (like why some particular bird migrates south for the winter, or why a particular person becomes schizophrenic), the answer lies in understanding the building blocks that make up those phenomena—and that those building blocks are ultimately genes. In this deterministic view, the proteins unleashed by genes "cause" or "control" behavior. Have the wrong version of a gene and, bam, you're guaranteed something awful, like being pathologically aggressive, or having schizophrenia. Everything is preordained from   conception.

3. Article by Robert Sapolsky from *Newsweek* April 10, 2000. Copyright © *Newsweek*. Reprinted with permission.

Yet hardly any genes actually work this way. Instead, genes and environment interact; nurture reinforces or retards nature. For example, research indicates that "having the gene for schizophrenia" means there is a 50 percent risk you'll develop the disease, rather than absolute certainty. The disease occurs only when you have a combination of schizophrenia-prone genes and schizophrenia-inducing experiences. A particular gene can have a different effect, depending on the environment. There is genetic vulnerability, but not inevitability.

The Primacy of Genes also assumes that genes act on their own. How do they know when to turn on and off the synthesis of particular proteins? If you view genes as autonomous, the answer is that they just know. No one tells a gene what to do; instead, the buck starts and stops there.

> *... genes don't independently determine when proteins are synthesized. They follow instructions originating somewhere else.*

However, that view is far from accurate too. Within the staggeringly long sequences of DNA, it turns out that only a tiny percentage of letters actually form the words that constitute genes and serve as code for proteins. More than 95 percent of DNA, instead, is "non-coding." Much of DNA simply constitutes on and off switches for regulating the activity of genes. It's like you have a 100-page book, and 95 of the pages are instructions and advice for reading the other five pages. Thus, genes don't independently determine when proteins are synthesized. They follow instructions originating somewhere else.

What regulates those switches? In some instances, chemical messengers from other parts of the cell. In other cases, messengers from other cells in the body (this is the way many hormones work). And, critically, in still other cases, genes are turned on or off by environmental factors. As a crude example, some carcinogens work by getting into cells, binding to one of those DNA switches and turning on genes that cause the uncontrolled growth that constitutes cancer. Or a mother rat licking and grooming her infant will initiate a cascade of events that eventually turns on genes related to growth in that child. Or the smell of a female in heat will activate genes in certain male primates related to reproduction. Or a miserably stressful day of final exams will activate genes in a typical college student that will suppress the immune system, often leading to a cold or worse.

You can't dissociate genes from the environment that turns genes on and off. And you can't dissociate the effects of genes from the environment in which proteins exert their effects. The study of genetics will never be so all encompassing as to gobble up every subject from medicine to sociology. Instead, the more science learns about genes, the more we will learn about the importance of the environment. That goes for real life, too: genes are essential but not the whole story.

Worm Offers the First Animal Genome[4]

By John Travis
Science News, December 12, 1998

Pity poor *Caenorhabditis briggsae*. Like most nematodes, this worm lives an uncelebrated life. Three decades ago, however, the millimeter-long worm had its big shot at fame-and came up short.

At that time, biologists were picking a worm species on which to concentrate their genetic and developmental studies. Sydney Brenner, who led the effort, had his eye on *C. briggsae*. But when the dust settled, a relative, *Caenorhabditis elegans*, had stolen the part.

Testifying to its subsequent rise to biological stardom, *C. elegans'* sinuous body graces the cover of the Dec. 11 *Science*. In the journal, investigators announce that they have essentially finished sequencing the worm's genome, making it the first multicellular organism whose full set of genes is known.

That feat thrusts *C. elegans* and the small, tightly knit community that studies it into the scientific spotlight. "It's fun," says nematode biologist Gary Ruvkun of the Massachusetts General Hospital in Boston. "The worm genome, being the first animal genome, becomes the brightest lighthouse for the navigation of all animal genomics."

Compared with viruses, bacteria, or yeast—the other organisms whose genomes have been sequenced—the microscopic *C. elegans* much more closely resembles complex animals, including people. It has a nervous system that includes a simple brain. It digests food, usually a steady diet of bacteria. It even reproduces sexually by fertilizing its eggs with sperm.

Easy to cultivate in the lab, the worm has a transparent body that simplifies the study of how a fertilized egg develops into an adult. Biologists have traced back to the egg the history of each of the nearly 1,000 cells in the adult animal. They've even mapped the connections between every nerve cell in the worm.

The sequencing of *C. elegans* DNA, led by scientists at the Sanger Centre in Cambridge, England, and the Washington University School of Medicine in St. Louis, began about 8 years ago amid skepticism that such a large genome could be deciphered. "There were certainly lots of people who thought we were foolish," says Robert Waterston, who directed the St. Louis effort.

The task turned out to be even more complex than expected. Early estimates suggested that *C. elegans'* six chromosomes had 6,000 or so genes, about half of them essential for life. The completed genome revealed nearly 20,000 genes, however. In comparison, the genome of the yeast Saccharomyces cerevisiae contains fewer than 7,000 genes (*SN*: 5/4/96, p. 278).

The overlap between the yeast and worm genomes suggests that there's a core of about 3,000 genes that are crucial to the workings of eukaryotic cells. Such cells, which include human cells, package most of their DNA inside a nucleus. The core genes encode proteins that play roles in basic activities such as DNA synthesis, the making of the cell's skeleton, the transport of proteins, and chemical signaling within the cell.

Curiously, in the worm, most of the essential genes cluster within the central regions of each chromosome. Noting that chromosome ends may evolve more rapidly, Waterston suggests that "evolution found a mechanism to tuck these genes away in a safe place." In contrast, he says, the chromosome arms may be "gene nurseries, or graveyards, or both."

The 17,000 or so additional genes within the worm, compared to yeast, should help biologists explain how multicellular animals differ from single-cell eukaryotes. Many of these genes, for example, encode proteins for cell-celladhesion and signaling between cells.

To identify all the nematode's genes, the researchers had to spell out a DNA sequence of 97 million bases, the chemical building blocks of DNA. A few small gaps remain, although the scientists are confident they have sequenced more than 99 percent of the worm's genes.

Waterston notes that the many short, repeating DNA sequences in *C. elegans'* genome, and their distribution, made sequencing it even more challenging in some ways than the human genome. The methods developed to analyze the worm genome are already speeding the human genome effort, he says.

The nematode's genome itself should offer insight into human biology. "Because *C. elegans* has so many genes shared with humans . . . we can figure out what they're doing in *C. elegans* and apply that to a whole myriad of human genetic issues," says Waterston.

Scientists also hope that knowledge of the *C. elegans* genome will help people deal with worms that cause human illnesses and agricultural problems. "It is important that the *C. elegans* genome project yields an improved understanding of other nematodes, so as to enable the development of control strategies to alleviate their effects on human populations," Mark Blaxter of the University of Edinburgh says in one of the six articles on the nematode in the Dec. 11 *Science*.

The worm genome marks a new era in evolutionary biology, one in which relationships between animals will be based on genomes rather than on fossils or a single gene. Already, scientists are reconsidering whether nematodes developed before or after the time when arthropods and vertebrates parted ways.

What's missing from the *C. elegans* genome may provide as much evolutionary insight as what's there. "What it looks like is that certain things have been deleted in the lineage leading to *C. elegans*," says Ruvkun. "Having genome sequences for a wide zoo of creatures will be very important." The fruit fly's complete genome is expected next, he notes.

> *The worm genome marks a new era in evolutionary biology, one in which relationships between animals will be based on genomes rather than on fossils or a single gene.*

Biologists caution that sequencing an animal's genome is but a first step toward a better understanding of the animal. More than half the newfound *C. elegans* genes have no identified function. "We have to fill in a considerable amount of what the gene products do inside the cells they work in," says Brenner, who now is at the Molecular Sciences Institute in Berkeley, Calif.

Researchers plan to inactivate every one of the worm's genes in an attempt to understand each one's role in development and adult life. Also on the drawing board are microchips covered with the worm's DNA, which will allow the simultaneous monitoring of the activity, or expression, of all of *C. elegans'* genes (*SN:3/8/97, p. 144*).

Biologists also propose to make movies of every gene's activity in the worm's development and life. To do this, they intend to fuse each of its genes to a jellyfish gene encoding a light-emitting molecule. "We would know exactly when and where each gene was expressed," says Waterston.

With all this attention on *C. elegans*, it's easy to forget *C. briggsae*. That worm did land a supporting role. Scientists are now sequencing parts of its genome for comparison. Why wasn't it chosen for stardom? It's just not as photogenic or easy to work with, notes Brenner.

Fruit Fly Genome Yields Data and a Validation[5]

BY ELIZABETH PENNISI
SCIENCE, FEBRUARY 25, 2000

The humble fruit fly has just soared to the top of the genome charts. Using an approach dismissed as unworkable a mere 2 years ago, a team of publicly and privately funded scientists announced last week that they had decoded more than 97% of the genome of Drosophila melanogaster. As with all genome projects, parts are missing: The team sequenced only gene-containing regions, and about 1600 gaps remain. Even so, Drosophila, which has been long studied by geneticists, is the largest creature ever to be sequenced, genomewise, and only the second multicellular organism. What makes this milestone especially noteworthy, however, is that it validates the controversial "shotgun" approach. As such, it could pave the way for a public-private effort to complete the human genome, said J. Craig Venter, president of Celera Genomics in Rockville, Maryland, the private half of the team.

The last two big successes of the genome project, the nematode (*Science,* 11 December 1998, pp. 1972, 2012) and human chromosome 22, recently published in *Nature* (*Science,* 24 September 1999, p. 2038), were both done using the "clone-by-clone" approach. This involves determining the order of the bases in a series of overlapping clones, whose locations on the chromosomes are known.

In May 1998, Venter stunned the genome community when he said he would tackle the human genome with the whole-genome shotgun approach that he had pioneered on microbial genomes (*Science,* 18 June 1999, p. 1906). To "shotgun" a genome researchers shred the entire genome into random pieces, sequence all the pieces, and then reassemble them in the correct order with the aid of a supercomputer. At the time, critics argued that Venter would be unable to put the millions of DNA fragments back together. As a test case, Venter teamed up with Gerald Rubin and the Berkeley Drosophila Genome Project to try the fruit fly.

The effort "worked better than anyone expected," Rubin reported at the annual meeting of the American Association for the Advancement of Science, which publishes *Science*. Geneticists and

---

5. Reprinted with permission from *Science Magazine.* Copyright (2000) American Association for the Advancement of Science.

molecular biologists are ecstatic. "[Venter and Rubin] have really pushed the envelope of what's possible," raved Daphne Preuss, a geneticist at the University of Chicago. Added geneticist Lawrence Goldstein of the University of California, San Diego: "The quality of what I saw was really exceptional."

One key to their success was an assembly program designed by Celera's Eugene Myers. In short order, the program was able to assemble the 120 million bases into 26 long stretches, or "scaffolds." Myers relied on existing genome maps to order these stretches. Still, the program left 1800 gaps, which Myers reduced to 1600 by adding sequence data from his academic collaborators. What's more, the shotgunned data matched already finished fly sequence quite well. Myers is confident that this approach will work on the far larger human genome. But skeptics are waiting to see how difficult the remaining gaps are to close, a task Rubin's team is taking on, before giving the thumbs-up.

Meanwhile, analyses so far suggest the fruit fly could have as many as 13,000 genes, half of whose functions are unknown, said Celera's Mark Adams. With the sequence in hand, Goldstein expects research to "catapult ahead." For Rubin, this achievement is sweet because everyone worked together well: "It has been one of the most pleasurable scientific experiences that I've had in my academic career."

Our Genes for Sale—Get 'em While They're Hot[6]

By Ellen Goodman
Boston Globe, March 8, 2000

I have to tip my cap to Donna MacLean. The British poet and waitress walked into a U.K. patent office and filed an application for a very unusual invention: "Myself." "It has taken 30 years of hard labour for me to discover and invent myself," she wrote, "and now I wish to protect my invention from unauthorised exploitation, genetic or otherwise." MacLean was motivated by a sense of humor and a sense of outrage. It seems she'd been reading about companies trying to patent genes and decided she'd better hang on to her own private property.

We are now close to deciphering all the human genes. Scientists are expected to complete sequencing the 3 billion units that make up our DNA by the end of the year.

Hers is more of a statement than a solution; it's better performance art than patent law. But she's onto something.

Most of these 3 billion bits are, I hate to say, pretty worthless, no matter how attractively they may be arranged in MacLean. We share 30 percent of our genes with lettuce.

But some genes may hold secrets that cure cancer, Alzheimer's and other scourges of our genetic inheritance. So the Human Genome Project is putting sequences into the public venue—the Internet—for all researchers as fast as it can.

Meanwhile, a handful of biotech companies like Celera, run by the flamboyant J. Craig Venter, are racing to stake a private claim on whole stretches of DNA. The president of Incyte has bragged that whoever wins the patenting race will become the "eBay for genes." They can offer them for commercial use to the highest bidder.

The gold rush for patents on every gene, whether it has a known use or not, has gotten the scientific community and at least one poet-waitress worried.

It's not just the philosophical questions—Who Owns My Body? Who Owns the Entire Genetic Repository of the Human Species?—that spook people. (Let's remember that eBay was just asked to auction off a human soul.) There is also a question about

---

the public good. What's the best, fairest, fastest way to go from mapping the human genome to curing human disease? There is something to be said for patenting. In theory, it gives private companies the incentives to get the medicine to the marketplace. Scientists may be satisfied with Nobel Prizes, but stockholders want returns.

On the other hand, patenting can inhibit progress by making it prohibitively expensive. The marketplace does a dicey job of regulating medicine. Just compare the falling cost of a Palm Pilot and rising cost of medicine.

Francis Collins, the head of the Human Genome Project, uses a toll booth analogy. "Medical research is a road to discovery," he says.

When can patenting help and when can it hurt? At the moment, Ellen Clayton, director of Vanderbilt's Genetics and Health Policy Center, expresses a view shared by many that "patents are being awarded for too little intellectual work." Consider the current feud over rights to a gene that may be important in AIDS therapy. A group of AIDS researchers discovered a promising use for CCR5.

But a biotech company that did little more than list the letters and number on the code already had a patent on its commercial development.

It was, rued researcher Robert Gallo, as if the biotech company said, "I found a fungus, therefore I should get credit for penicillin." The questions being raised now are: Should the people—or even the computer—doing the relatively routine work of identifying a gene get to be the landlord charging whatever the market will bear? Or should that property go to the person who figures out what to do with it? Francis Collins, the head of the Human Genome Project, uses a toll booth analogy. "Medical research is a road to discovery," he says.

You start out on a bumpy path with a lot of ruts, wrong turns and dead ends. You get along far enough until you see a clear way to turn knowledge into a product.

"At that point, it's OK to have a toll. You need to do the upkeep on the road. The patent maintains the road." But if the tolls start piling up too early, too often and too expensively, nobody will take that road. "The tendency in biotechnology has been to put the toll booths earlier and earlier," says Collins. Business is hogging the freeway.

Patent law and the Patent Office were set up long before anyone heard of DNA. Now the whole blueprint is in play.

In France, you can't patent genes at all. In Britain, our gal Donna is trying to get a poetic grip on "myself." Before someone corners the American market on chromosomes, we better protect public access to a nice, clean, affordable set of genes.

"We share 30 percent of our genes with lettuce."

II.

Genetic Testing: Identification and Diagnosis

Editor's Introduction

Mark Twain's 1894 novella *Pudd'nhead Wilson* tells the story of a small town dazzled by a lawyer's use of fingerprints to solve a local crime and prosecute the suspect. Several years after its publication, fingerprinting revolutionized the field of crime fighting, its effectiveness based on the uniqueness of each person's prints. Until 1953 it was the only known physical evidence of individual singularity, the only definitive means of distinguishing one person from another. Then, in 1953, doctors James Watson, Francis Crick, and Maurice Wilkins discovered the molecular structure of DNA, the long, double helix pictured on the cover of this book that contains the code for each individual's genetic composition. While a human fingerprint can be altered or obliterated when fingertips are burned by fire or acid, DNA is indelible and unmistakable. It also provides more extensive information about a person than a fingerprint, from the most superficial characteristics like hair and eye color, to significant facts about one's health and physiology, like a predisposition to cancer. In recent years the use of DNA testing by both doctors and crime fighters has steadily increased, as it is employed to diagnose and, in some cases, determine a course of treatment for medical conditions, to identify suspects, and to convict criminals. Nevertheless, many fear the consequences of these tests. Some worry that the very personal information revealed in our DNA may be used to discriminate against us, while others fear the psychological harm should we find ourselves carriers of a gene for a currently untreatable and debilitating disease, such as Huntington's. Section II explores these important issues from a variety of perspectives that look forward with both eagerness and trepidation to the future of genetic testing.

The first article in this section, a piece from *Reason* called "Unlocking the Cells," written by Ronald Bailey, discusses a new technique called DNA fingerprinting that may be used in both local and federal law enforcement to convict the guilty and exonerate the innocent. After briefly describing England's innovative DNA identification network, which served as a model for the FBI's National DNA Index System, Bailey cites examples of effective uses of DNA fingerprinting in the U.S. since 1998. Among these examples is the Innocence Project, founded by Peter Neufeld and Barry Scheck, which has helped to free mumerous inmates convicted before DNA testing technologies were available. Despite fears that such testing could be perceived as invasive, Baily argues for more widespread use of DNA fingerprinting, claiming it would provide a "check on the government" and a means of ensuring that justice is served.

The next article in this section explores the debate over genetic testing in another context, that of adoption. In "DNA Tests Cast Shadow on Adoption," Richard Saltus of the *Boston Globe* reports on the controversy surrounding the use of DNA in adoption proceedings, a practice that many consider invasive and that could potentially hurt a child's chances of being adopted. Child advocates fear that a couple's desire to have "the perfect family" will prompt them to discriminate against children whose DNA suggests future health problems and cause them to reject potential adoptees, thereby increasing the number of children in foster care. As Saltus reports, however, others argue that such testing could benefit children by providing their adoptive parents with knowledge of genetic abnormalities, thereby enabling doctors to begin immediate treatment for genetic syndromes or other conditions. This issue, one of the most contentious in this area of biotechnology, is not likely to be resolved soon, as adoptive parents are expected to increase pressure on agencies to employ the most advanced methods available to investigate potential adoptees.

In a further discussion of the use of genetic testing in family planning, Kristi Coale, writing for *Salon*, points to what she considers the dubious history of the American Eugenics Movement as a cautionary tale for today. Her article "Playing God" examines tales of parents who considered or sought sterilization procedures after state-of-the-art tests indicated—often incorrectly—that they were mentally or physiologically inferior or that they carried genes for specific illnesses. Coale also questions the notion that "biology is destiny" and compares the use of today's prenatal genetic testing with past techniques of determining potential criminality by examining the shape of a child's head. As this article demonstrates, whether the discussion is about prenatal genetic testing, in utero surgery, or the attempt to purchase the eggs of fashion models on the Internet, the issue of control is primary. The question Coale asks is, to what extent are we prepared to employ the tools of science toward the exercise of that control over human life?

Unlocking the Cells[1]

By Ronald Bailey
REASON, January 2000

The impeachment of President Clinton underscored the growing importance of DNA evidence in criminal investigations. If it weren't for the telltale stain on a certain blue dress, the president might still be insisting that he "never had sexual relations with that woman, Ms. Lewinsky."

DNA testing was first used in Britain in 1986 to prosecute serial rapist and murderer Colin Pitchfork. Today, DNA testing is regularly used to convict criminals, much as fingerprints have been for many years. "No other form of evidence for identifying human beings has gone through such a rigorous scientific and legal validation as DNA has," says Christopher Asplen, executive director of the National Commission on the Future of DNA Evidence, a panel of expert advisers set up by the U.S. Department of Justice. "Now it's the most reliable evidence we've got." DNA testing is a powerful way to identify people because nearly every human cell contains it, and each person's DNA is unique (except in the case of identical twins). In 1998, the Federal Bureau of Investigation created the National DNA Index System (NDIS), which links the DNA databases of 18 states so far. Eventually, all 50 states are expected to participate in the NDIS. These databases currently contain the genetic profiles of some 210,000 criminals and are expanding rapidly.

The profiles are based on DNA samples collected from people who have been convicted of murder, manslaughter, rape, or aggravated assault. Some states, such as Virginia, require that all convicted felons provide DNA samples for profiling. The databases are far from complete. Paul Ferrara, director of the Virginia Division of Forensic Science, estimates that the DNA of 1 million felons nationally should have been collected but has not been and that half a million samples that have been collected are still not profiled.

Despite these shortcomings, the databases have dramatically proven their value, solving scores of old murder and rape cases by matching DNA evidence from those crimes to DNA profiles. Florida claims to have made some 200 "cold hits" using the databases,

---

1. Reprinted with permission, from the January 2000 issue of *Reason Magazine*. Copyright © 2000 by the Reason Foundation.

and Virginia reports 78. A "cold hit" occurs when police who have no leads find a suspect by checking the DNA from a crime scene against the DNA profiles in the databases. Great Britain was an early innovator in DNA profiling, and British police claim to solve 300 to 400 crimes per week using DNA databases. DNA databases are effective because many criminals make a career of crime: In two studies, one in 1991 and another in 1995, political scientist John Dilulio reported that, based on interviews with prison inmates, they had committed an average of 12 crimes in addition to the ones for which they were caught and convicted.

Various types of DNA fingerprinting have been developed since the mid-1980s, but state-of-the-art DNA profiling depends on what are called "short tandem repeat" polymorphisms (STRs). STRs are segments of DNA that show considerable variation between individuals. Criminal investigators have adopted a standard using 13 STR core loci for identifying genetic differences between people. These loci are not genes but areas of "junk" DNA found in all human beings. Testing one locus gives a 1-in-500 chance that a particular sample of DNA came from a particular individual. Testing all 13 loci changes the odds to 1 in 82 billion. At $50 a profile, STR profiling is significantly cheaper than earlier technologies.

> *Great Britain was an early innovator in DNA profiling, and British police claim to solve 300 to 400 crimes per week using DNA databases.*

DNA testing is also a powerful tool for exonerating people who have been imprisoned for crimes they didn't commit. In September, the National Commission on the Future of DNA Evidence released a report, *Postconviction DNA Testing: Recommendations for Handling Requests*, addressing such cases. "Commission members have an absolute sense that every single day that some innocent person sits in jail is too long," says the DOJ's Asplen.

The commission's report outlines a process by which prosecutors, defense attorneys, and judges can evaluate requests by inmates for postconviction DNA analysis based on the likelihood that such an analysis, had it been available at the time of conviction, would have changed the verdict. The report recommends that prosecutors and judges, if they can, waive the time limits on motions for a new trial based on newly discovered evidence of innocence. States established such time limits because evidence deteriorates, memories fade, witnesses die or move, and courts should not waste scarce resources on retrying old cases. Moreover, many states mandate the destruction of evidence after a certain period of time.

"The biggest problem with the commission's recommendations is that they are only recommendations," says defense attorney Peter Neufeld. "Prosecutors don't have to follow them if they don't want

A Step-By-Step DNA Primer

The human body has 3 trillion cells. All except red blood cells contain DNA—deoxyribonucleic acid, the chemical that stores each person's genetic code. Even after being multiplied millions of times, DNA is invisible to the naked eye. Tests duplicate and isolate 13 specific sites along a DNA strand, a coiled helix that in just one cell stretches 5 feet. These sites identify an individual. Here's how it's done:

1. Technicians start with a fluid sample, usually blood or semen, collected from a crime scene or victim. Scientists are experimenting with extracting DNA from skin.

2. At least 100 cells from one source are needed for a DNA test. The cells are put in a chemical solution, heated and cracked open to release the DNA.

3. Next comes something called PCR—polymerase chain reaction. Technicians apply a synthetic primer that acts like a copying machine, stimulating duplication of the 13 comparison areas millions of times over.

4. High voltage separates the 13 DNA sites by size in a sequencer hooked to a computer. This machine—the size of a dorm room refrigerator—contains a laser that hits the DNA, making it glow. At least 100 cells from one source are needed for a DNA test. The cells are put in a chemical solution, heated and cracked open to release the DNA.

5. The computer reads the glowing material and prints out a graph of numbers next to lines forming peaks and valleys. These lines represent the lengths of the pairs of DNA markers—called alleles—at those 13 sites. The FBI has assigned numbers to each possible configuration of the alleles at those sites. The resulting computer printout is a bar code with numbers. Everyone's is unique.

6. Once the computer-generated pattern of an individual's DNA is established, technicians repeat the two-day process to verify the results.

By Aram Adourian, courtesy of Whitehead Institute for Biomedical Research.

to." Neufeld and Barry Scheck, the DNA experts for O.J. Simpson's defense team, founded the Innocence Project in 1991 at Yeshiva University's Cardozo School of Law. The Innocence Project focuses on using DNA evidence to help inmates exonerate themselves. So far 65 inmates nationwide have been freed using DNA evidence, usually in cases where DNA testing technologies were not available at the time of the original trial.

Neufeld bases his concerns about whether prosecutors will follow the commission's recommendations on his experience that in most of the cases taken on by the Innocence Project, prosecutors refused to consent to DNA testing. Neufeld estimates that thousands of prisoners might be exonerated if DNA evidence were available for testing. But there is no biological evidence to evaluate in 70 percent of the cases initially reviewed by the Innocence Project. Nevertheless, Neufeld says they have a backlog of 1,000 cases, of which only 200 are being actively pursued.

> *So far 65 inmates nationwide have been freed using DNA evidence, usually in cases where DNA testing technologies were not available at the time of the original trial.*

Right now only two states, Illinois and New York, have laws giving prisoners the right to postconviction DNA analyses. Neufeld and Scheck propose that the federal government or every state adopt legislation that would permit postconviction DNA testing that would be paid for by the government.

Also, they want laws enacted that would allow access to evidence even for inmates who do not meet the threshold criteria for states to pay for DNA testing but who are willing to pay for the DNA tests themselves. Also, they want no time limits on new trials in which new DNA evidence could exonerate an inmate. Neufeld and Scheck recommend that biological evidence from a crime scene be preserved at least as long as an inmate convicted for that crime on the basis of that evidence is in prison—a point on which Virginia's Ferrara concurs.

As DNA testing improves, other questions will be raised. Now only convicted felons must submit to DNA testing. In the future, will police be able to require suspects and arrestees to provide DNA samples for matching against crime scene evidence and DNA databanks? Ferrara analogizes this use of DNA testing to fingerprinting. Today, arrestees must provide the police with their fingerprints. Someday they might have to hand over blood or other tissue for DNA sampling. Soon, Asplen notes, DNA testing technology will be able to use the cells left behind by ordinary fingerprints for DNA profiling.

Since it doesn't test for genes, STR profiling can only identify people and does not provide any genetic information that might be of interest to, say, health insurers. But even after the STR profiles are obtained, Virginia and other states keep the biological samples taken from felons. It is not too hard to imagine future researchers trying to gain access to those samples in order to prospect for genes that predispose people to violent or otherwise antisocial behavior. Also, British investigators are beginning go beyond simple STR identifiers to look for DNA markers in genes for eye color, hair color, and race that would help them construct fuller physical profiles of suspects.

Britain and France often use "voluntary" mass screenings to find criminals in a community. For example, in 1996 the murderer of a 15-year-old girl was caught through a mass screening of 2,000 local men in Cardiff, Wales. For those who say it can't happen here, Neufeld reports that a Florida investigator recently told him of a murder case in which 250 men were asked to provide DNA samples voluntarily. The investigator told Neufeld, "Not a single man we asked refused to provide a sample." The murderer was caught—not through DNA screening but when he tried to kill another victim.

Today, arrestees must provide the police with their fingerprints. Someday they might have to hand over blood or other tissue for DNA sampling.

In the science fiction movie *Gattaca*, which portrays a dystopian world of genetic haves and have-nots, the hero is nearly caught by the gene police when he leaves a stray eyelash in the wrong place. This scenario is not so far-fetched, since everyone is constantly shedding DNA-containing material such a dry skin cells, hair, and saliva. Improved DNA testing techniques could use this type of everyday DNA evidence to keep track of people.

Clearly, DNA testing is an increasingly powerful forensic tool, and we should be on constant guard for potential abuse of it by the authorities. But the exoneration of convicts based on DNA evidence shows that the technology can also be used as a check on government, and proposals for public financing of these tests make sense. Indeed, the federal and state governments should be eager to pay for DNA testing and analysis to be sure that no innocent person has been wrongfully imprisoned. After all, if the government isn't about rendering justice, what is it about?

DNA Tests Cast Shadow on Adoption[2]

Biotech conference addresses patients' rights, ethics

By Richard Saltus
Boston Globe, March 27, 2000

The gurgling infant girl awaiting adoption seems healthy now.

But what surprises may be encoded in her DNA? Are there flawed genes that predispose her to serious mental or physical health problems—perhaps soon, perhaps years from now—that would not only cloud her life, but put a severe burden on her adoptive parents?

DNA testing is growing in power to predict future health or disease, and with the Human Genome Project rushing toward its end, all sorts of genetic forecasting will increase.

In fact, the speed with which science is gaining the ability to predict who will fall victim to debilitating diseases is outpacing our power to cure or even treat those same ailments.

The dilemmas posed by such issues as who should be tested, who should be told, and who should be protected from involuntarily submitting or releasing information about their genetic makeup are among the thorny ethical issues being discussed among the thousands of biotechnology leaders convening this week in Boston for the Bio2000 convention.

Although genetic testing for predisposed medical conditions could invade every facet of human life, adoption specialists say it is already a rising concern among prospective adoptees.

"We've had a number of calls asking for just this kind of thing," said Karen Eanet, a genetic counselor at Greater Baltimore Medical Center. "They ask for some things we can't do and things we can do but wouldn't recommend."

The issue of predictive genetic testing of adoptees raises difficult moral questions. In whose interest would it be to know about medical problems ahead of time? What if the test detects conditions that aren't treatable? Does testing invade the child's right to privacy?

Courts have ruled that adoption agencies have to provide all reasonable medical history information. That hasn't been legally inter-

---

preted to mean that agencies have to search for more information, such as a genetic test might provide.

But with hundreds of tests that can diagnose rare conditions and others on the horizon that could identify predispositions to more common diseases far in advance of symptoms, there could be incentives for couples to ask for genetic testing of a child before they adopt.

The dilemma pits the interests of some parents, those who are worried about hidden health problems, against the interests of children for whom agencies are seeking families. And, said specialists at an unusual meeting on adoption and genetic testing earlier this month, it raises the question of whether genetic technology has created a demand for "the perfect baby" or "the perfect family," so that children with some abnormalities would be less likely to find homes.

"If we do these tests, it sounds like we'll have a huge pool of unadopted children rotting in the foster care system," said Susan Harris, a regional manager at Children's Services of Roxbury.

"Genetic testing has the possibility of putting some children in the back of the line, and that really frightens me," said Susan Soon-Keum Cox, an authority on international adoption who was adopted from Korea in 1956.

Harris and Cox participated in a meeting earlier this month at the University of Massachusetts in Boston that brought together specialists from many fields to discuss genetic testing.

> *"If we do these tests, it sounds like we'll have a huge pool of unadopted children rotting in the foster care system."*— **Susan Harris, Children's Services of Roxbury**

"My understanding is that right now there aren't a lot of requests for this type of testing, but as the tests multiply and become more accurate, this problem will get more acute," said Aytan Y. Bellin, counsel for a New York State task force weighing this and other bioethical issues.

Janet Farrell Smith, the conference's organizer and a professor of philosophy at UMass, said one incentive for genetic testing is a lack of medical records in many cases, particularly in foreign adoption. In such cases, both the child and the adoptive parents could have an interest in genetic testing that would reveal a proneness to health problems that runs in the birth family.

A major question about preadoption gene testing, said Smith, is: "Will you violate the child's privacy rights and risk genetic discrimination in the future?"

The conference brought together specialists from such disparate fields as medical ethics, adoption services, law, and medicine.

Some argued that the main purpose of adoption is to find homes for children and that genetic testing could harm their chances.

Bellin referred to two cases in which couples considering adoption sought genetic testing to reveal whether an adoptee would develop Huntington's disease, a fatal, inherited disorder. In both cases, specialists refused to perform the test.

Genetic tests already can detect rare but serious disorders in infants and can find single malfunctioning genes that cause diseases such as Duchenne muscular dystrophy and cystic fibrosis.

More problematically, DNA tests also can reveal the presence of genes that increase a person's chance of developing heart disease, Alzheimer's disease, and inherited cancers. But these tests can't predict with certainty, because environmental factors also help determine whether the disease develops.

Genetic tests already can detect rare but serious disorders in infants and can find single malfunctioning genes that cause diseases such as Duchenne muscular dystrophy and cystic fibrosis.

Leonard Glantz, a lawyer and bioethicist at the Boston University School of Public Health, argued that testing for adult-onset diseases would violate the child's right to privacy. And, even though it might benefit anxious parents who are worried about an adopted child's future health risks, "My position is you don't treat parent anxiety by testing their children for adult-onset diseases," he said.

Dr. Laurie Demmer, a medical geneticist at UMass Memorial Health Care in Worcester, said she often consults with parents considering adoption and wanting to know whether the child might develop a life-threatening disease such as cystic fibrosis or might have a chromosome defect that could cause mental retardation.

At other times, the child may have some abnormalities, such as a cleft lip, that could be part of a genetic syndrome. That's where a genetic test might add information that will help immediately in treating the child.

But, she said, she also has requests for tests she won't do. "I spend a lot of time explaining why we don't test children for late-onset disorders" such as Huntington's disease or to detect genes they may carry that are harmless to them but could be passed on to a child. "We like to wait until they're old enough to make the decision themselves," Demmer said.

Perhaps the most vexing dilemma the participants discussed was whether prospective adoptees should be tested for a disease such as Duchenne muscular dystrophy. This fatal genetic disorder often isn't diagnosed until the child is several years old and begins to show muscle weakness. Patients become progressively disabled and die in their teens or early 20s.

In this case, genetic testing of an infant wouldn't lead to any medical benefit for the child, since there is no treatment. But parents may well feel that they're not equipped to deal, financially or emotionally, with such a devastating condition.

Elizabeth Bartholet, a professor at Harvard Law School and an author on adoption and reproductive issues, said she feels that prospective adoptive parents "have some right to know about extreme problems they don't feel ready to take on." Others, however, argued against any attempt to match children's needs with parents' desires or ability to take care of them. Nobody, they argued, is prepared for a child with severe medical problems.

The conference was not aimed at creating a position on testing. But speakers noted that the American Society of Human Genetics and the American College of Human Genetics have taken positions against testing except to detect conditions for which treatment would have immediate benefits.

It remains to be seen how the issue will play out. Anita Allen, a professor of law and philosophy at the University of Pennsylvania, said, "I think the public adoption sector is extremely reluctant to move in the direction of genetic testing."

But in private adoptions, said Allen, "agencies have to depend more on the values of consumers." "I think there may be more pressures" to perform testing and more ambiguity about the decisions, she said.

Playing God[3]

Scary eugenics documents from the turn of the century shine a disturbing light on ethical dilemmas raised by genetic testing.

BY KRISTI COALE
SALON, NOVEMBER 17, 1999

Something about Vivian Buck troubled a Red Cross aide, though the relief organization worker couldn't quite put her finger on it. All she could say about the 7-month-old Vivian was that there was a "look" about her that was "not quite normal." This observation was the missing piece in a puzzle officials at Cold Spring Harbor Laboratory's Eugenics Record Office were trying to solve: They wanted to prove that feeblemindedness was a trait passed from parent to offspring.

Already the laboratory had IQ test scores for Vivian's mother Carrie and her grandmother Emma which found the women to be "morons." Adding the "data" about Vivian's looks to the mix was enough to establish that three generations of the Buck family were of low intellect. These facts became the basis of a landmark 1927 Supreme Court decision that allowed states to forcibly sterilize people who carried "hereditary defects." Carrie Buck was forcibly sterilized, and by the mid-1930s, about 20,000 people in the United States met the same fate under similar laws.

Vivian Buck's story, along with various state sterilization laws, are among the artifacts that will soon be on the Web as part of a digital image archive chronicling a dark chapter in U.S. history—the American Eugenics Movement. The movement, which began in 1904, was a government-sponsored social engineering project which sought to improve the human species by encouraging "fit" people to marry and procreate while sterilizing and prohibiting unions between the "unfit."

The Image Archive on the American Eugenics Movement is expected to go online in January 2000. Judging from a preview, it's a pretty powerful site, featuring a collection of troubling documents and pictures. There are photos of men arranged as if in a police line-up, which purport to show correlations between the size and shape of one's head and one's intelligence; there is a photo of a

---

young boy just out of diapers who was identified as a likely poten-tial criminal—a determination based on the shape of his face. There are family trees which track alcoholism and idiocy across the generations; and there are photos of the "fittest families"—who apparently evidenced no undesirable traits.

Up to now, the materials of the eugenics archive, which had been dispersed among several institutions including the American Philosophical Society in Philadelphia, have remained an obscure body of research, accessible only to scholars. By granting broad access to the archive over the Internet, David Micklos, director of

---

... I was struck by a disquieting common ground shared by eugenics and today's prenatal genetic testing: a belief that biology is destiny and that science alone can help us overcome it.

Cold Spring Harbor's DNA Learning Center and chief architect of the eugenics archive, hopes to encourage students and the general public to make a connection between what happened in the early 1900s and events in genetic research that are grabbing headlines today—a connection that could provide an ethical context for some agonizing decisions we face in our personal lives and in society. Clicking through some of the shocking images and articles of the exhibit, I was struck by a disquieting common ground shared by eugenics and today's prenatal genetic testing: a belief that biology is destiny and that science alone can help us overcome it.

The goal of the eugenics movement was to create the best society possible. Eugenics relied on the state-of-the-art genetics of its day, although the vast body of information that became the basis for sterilization, marriage and immigration laws aimed at weeding out those with "bad heredity" was based largely on anecdotal information. Field researchers at places like the Eugenics Record Office collected hereditary data through house-to-house surveys and the study of records of prisons, hospitals and institutions for the deaf, blind, insane and the mentally disabled. Today we have a lot more accurate information—scientific data about the structure of DNA and its proteins, which paints a detailed picture of how traits are inherited. And many of us—believing that technologies like prenatal genetic screening yield results that are reliable enough for us to use as the basis for life-altering decisions—are using this information to create the best children possible.

If you scoff at the notion that a child's future academic prowess was once thought to be determined by the fact that his or her grandfather was a drunkard, then consider that today we don't even flinch at the idea of controlling our own procreation. In fact, we expect to pay a lot of money to engineer the "perfect baby." Witness the advertisement placed in the student newspapers of Ivy League schools by an infertile San Diego couple last year. The ad offered $50,000 to college-aged women in exchange for their eggs. The couple, by selecting to advertise only to the Ivy League, was clearly looking for a particular kind of woman's eggs, and they were very specific: Potential donors had to be at least 5-foot-10, possess a combined SAT score of 1400 or higher, and show some athletic prowess.

Then there's the more recent case of Ron Harris, the Arabian horse breeder and fashion photographer who reportedly opened up a Web site for the purpose of auctioning off eggs and sperm of fashion models to the highest bidders. The price of admission to this little exercise is a fee of 20 percent of a the final bid on egg or sperm, and for that, you have the comfort of knowing that, "our striving reflects the determination to pass every advantage possible along to our descendants."

"I guess the question I ask about this is why buy such a specific egg," posits Paul Lombardo, professor at the Institute of Law, Psychiatry and Public Policy at the University of Virginia and a leading scholar of the Vivian Buck case. "I understand that couples are infertile, but [through ads like this], they're not just looking to have *a* baby, they're looking to have a *special* baby with special features they've picked out."

Why buy a specific egg? Don't we want the best for our children? Developments in reproductive and genetic science coupled with the age-old desire to give our children every advantage possible has led us to a point where we feel we are more responsible for our offspring at an earlier age—even before we conceive them. Of course, it's common knowledge that smoking and drinking alcohol can harm a developing baby, and a lot of women are aware that taking folic acid before and during a pregnancy decreases the chances of birth defects; these are the eat-your-broccoli type measures that any woman can take. But where we begin to cross the line from common-sense health choices to scientific control comes with procedures like prenatal genetic tests.

Prenatal genetic screening is designed to look for specific diseases such as Down syndrome and is recommended to women whom doctors determine to be "at risk." Just who is "at risk" is a decision that is made on based on results of preliminary blood tests and anecdotal and demographic information a pregnant woman gives her doctor. A woman's race, age and the diseases that have existed in her family and that of her mate's are among the facts used to determine the risks of producing a baby with a disease or disability. For example, risks for certain inherited diseases like Tay Sachs or sickle-cell anemia vary depending upon the race of a woman and her mate. So if a woman or her mate are Jewish, then she is likely to be screened for Tay Sachs.

The discussion between doctor and patient regarding prenatal tests often start off with the open-ended question: How do you feel about genetic testing? In the abstract, who could quibble with having more information about her developing baby? But when you're pregnant, wearing nothing but a hospital gown, and lying with legs agape on an examination table, this question can stir panic. That panic comes from the implicit message delivered by doctors when they advise patients to have these tests: Should the baby turn out to carry a disease or defect, intervention—in the form of in utero surgery or even abortion—is often advised.

In the abstract, who could quibble with having more information about her developing baby?

With prenatal genetic testing comes a small chance of triggering a miscarriage. At the same time, certain results, like tests for cystic fibrosis, can generate more questions than answers. Cystic fibrosis, a fatal, inherited illness where the body produces large amounts of abnormally thick mucus that accumulates in the lungs and intestines, is known to biologists as a single-gene disease. This means that having the gene would mean that someone would have the disease. But someone who carries the gene for the disease might never show signs of cystic fibrosis. That's because the genes themselves don't bring on the disease, says Garland Allen, a professor of biology at the Washington University St. Louis. The expression of the genes depend on other factors such as environmental triggers, Allen says.

Such ambiguities are not generally part of the discussion when doctors steer their patients toward having these tests. If you waver when asked the general, "How do you feel about testing?" question, then the doctor might ask, "What if," as in, "What if the baby has Down syndrome?" And if you're still waffling, and if you happen to have a child already, you might be asked, "What about your child? How will a Down baby affect your family?" This conversation hits a pregnant woman in her most vulnerable spot—her heartfelt concern for the health of her developing baby and the overall well-being of her family. And some doctors make their opinions about this clear: To undergo the tests is to fulfill a duty to do all that is possible to ensure a baby is healthy; to refuse the tests is to shirk responsibility.

What may feel like a coercive atmosphere to some pregnant women certainly doesn't compare to the collusive agenda of the eugenics movement, but it has similarly insidious consequences. The pressure to rely on science today is brought to bear on one patient at a time, and with the intent of ensuring the health of both mother and child, while the eugenics era was marked by a country-wide belief that everyone had a responsibility to do all that was possible to improve society, says Cold Spring Harbor's Micklos.

Groups like the American Eugenics Society cropped up all over the country after 1910, developing campaigns and sponsoring state fair exhibits to raise awareness of the importance of eugenics. The archive shows photos of billboard-type advisories in which flashing lights called attention to "facts" such as: Every 15 seconds $100 of public money goes to the care of "persons with bad heredity," and that every seven and a half minutes, "a high-grade person is born in the United States . . . with the ability to do creative work and be fit for leadership. About four percent of all Americans come within this class."

> *One irony today is that technology like genetic screening is being sold to us as a way of making our lives easier by reducing our chances of having to care for disabled children.*

One irony today is that technology like genetic screening is being sold to us as a way of making our lives easier by reducing our chances of having to care for disabled children. Yet one need only to look at the subject lines of postings to chat boards such as Parentsoup.com's Genetic Tests and Complications to understand the stress that these tests can cause. "One in 31 [spina bifida or other neural tube defect test] result and scared," posts one woman. "Low [spina bifida] at 17 weeks and scared to death—help!" writes another. Those yet to take tests or receive results, often ask "What if it's bad news?"

And this begs the question, how reliable are these technologies and what are the risks involved? There's the recent story of Nancy Seeger, the Chicago-based writer and artist who found through genetic tests that she was at increased risk for developing breast and ovarian cancer. Seeger looked at her family history with cancer—her mother and aunt died of breast cancer—and, with the advice of doctors, opted to have her ovaries removed. At the time of the surgery, Seeger donated some of her blood for study at a university hospital. Eight months later, doctors studying Seeger's blood found that she did not carry mutation of the gene for breast and ovarian cancer. The company that administered her genetic tests made a mistake.

The truth about modern genetic science is that the very information used as the basis for life-altering decisions like having your ovaries removed or life-ending decisions like aborting a cystic-fibrosis-carrying baby is not always conclusive. This is not to say that all genetic testing is bad. Despite the confusion and stress they cause, these tests and the medical interventions they enable have helped countless parents have healthy babies and others to avert a cruel illness. Instead, the stories of Nancy Seeger and the women on Parentsoup.com serve as cautionary tales of what we can lose in relying on science alone to make crucial decisions.

And therein lies the biggest lesson of the eugenics archive. The movement reached its zenith during the years that Cold Spring Harbor Laboratory operated the Eugenics Record Office, between 1910 and 1940. But by 1940, the laboratory shut down the Eugenics Record Office because the science that was used to make the various laws was discredited. For evidence of these shortcomings, one need look no further than the extensive records of the case of Vivian Buck. Among the images of the IQ tests and the observations of 7-month-old Vivian is an artifact that came a few years after the 1927 Supreme Court decision: Vivian's grade school report card. This record shows that Vivian's teachers found her to be bright. She had a solid B average, which proves that she was far from the imbecile that the high court found her to be.

Certainly, science and technology have ways of making our lives better. But they also make our lives harder, simply by giving us more options—which sometimes lead to hard choices. So it's inspiring to see the Internet come to the rescue with something like the eugenics archive—to put us in touch with history and force us to think about where science has taken us and where we need to go next.

III.

Reprogramming the Human Body

Editor's Introduction

I t is easy to sympathize with those individuals who learn through genetic testing that they carry a gene for a serious, debilitating illness, such as Huntington's or Alzheimer's disease, for which there is currently no cure. While they may not yet display symptoms, they could develop them at any time in their lives—or not at all. Living with the knowledge of their predisposition for such an affliction could be like living with a time bomb inside of their bodies, one that may explode at any moment. It is for people such as these that experts in biotechnology continue to develop the techniques that are the subject of Section III. Few experiences frustrate physicians more than having the ability to diagnose a serious, even fatal illness with no means to cure it, but in the last decade, that situation has begun to change. Many of the new and controversial treatments discussed in this section, such as stem cell transplants and gene therapy, have the potential to transform medical science and end the fear and helplessness that often accompany an unfavorable medical diagnosis.

The first article in Section III, "Capturing the Promise of Youth," written by Gretchen Vogel for *Science,* describes that journal's choice for 1999 Breakthrough of the Year: the medical applications of embryonic and adult stem cell transplants. Stem cells are cells at their most immature stage, the point at which they possess the greatest potential for development into any of the cells in the body. Vogel explains how research presently performed on mice suggests the possibility of treating a number of human ailments using stem cells, including growing transplantable organs in the laboratory, repairing nerve damage, restoring movement to paralyzed limbs, and reinvigorating weak or diseased muscles, hearts, and brains. Although debates continue in Europe and the United States on the ethics of using embryonic and fetal stem cells, Vogel reports that scientists have great confidence in their newfound power to "manipulate a cell's destiny."

The next article, "Petri Dish Politics," written by Ronald Bailey for *Reason*, reviews both sides of the debate mentioned by Vogel in the previous piece. After listing the latest developments in the field of biomedicine—including gene therapies attempted in mice to enhance intelligence and long-term memory, the creation of artificial chromosomes into which "genetic upgrades" might be inserted, "gene based" drugs to fight cancer, and a "biochip" to diagnose diseases—Bailey discusses the opposition by political conservatives to these and other innovative techniques, such as embryonic stem cell research. According to Bailey, these individuals, often allied with pro-life groups, are troubled by what they see as attempts by the biomedical community to eliminate suffering and death, which they view as essential experienced of human life. As he argues against this position, Bailey accuses conservatives of pressuring various members of Congress into refusing to fund important projects by the U.S. Department of Health and Human Services that could produce research, such as investigations into adult stem cells, of which they would approve. Ultimately, Bailey says, it comes down to a disagreement between those who fear we will abuse the tools of biomedicine to control the forces of life and death and those who believe we should accept and apply the knowledge we have while trusting our freedom to choose wisely.

Sherwin B. Nuland, writing for the *Wall Street Journal*, offers another perspective on this debate in his article "Immortality and Its Discontents." Taking a somewhat negative view of the kinds of innovations discussed in other articles in this section, Nuland points out what he considers the "serious problems" that will accompany these attempts to cure all diseases and prolong human life. He declares that only a "miracle" will prevent the aged from developing debilitating conditions, such as dementia, arthritis, and natural human frailty, as well as afflictions our species has yet to encounter. Nuland suggests that we encourage the elderly to prolong and improve the quality of their lives through an exercise regimen, rather than through procedures that would alter their genetic makeup.

The final article in this section, "A Thriving Pioneer of Gene Tests" by Marlene Cimons of the *Los Angeles Times*, reports on one young girl's successful treatment with gene therapy. Fourteen-year-old Ashanthi DeSilva, the first person ever to receive the controversial procedure, may not be cured of her illness (a rare disease afflicting her immune system), but the healthy genes injected into her bloodstream continue to reproduce, and there is hope that the stem cells they are planning to inject into her bone marrow will finish what the gene therapy started. As Cimons explains, while many scientists believe that Ashanthi proves the effectiveness of gene therapy, others consider her case merely anecdotal and therefore inconclusive.

Capturing the Promise of Youth[1]

By Gretchen Vogel
Science, December 17, 1999

Old age may have wisdom, but it will always envy youth for its potential. The same might be said for cells, for a young cell's most important trait is its ability to choose among many possible fates and become, say, a neuron passing electrical signals or a blood cell carrying oxygen. Late last year, in a technological breakthrough that triggered a burst of research and a whirlwind of ethical debate, two teams of researchers announced that they had managed to prolong the moment of cellular youth. They kept embryonic and fetal human cells at their maximum potential, ready to be steered into becoming any cell in the body.

Building on that achievement, in 1999 developmental biologists and biomedical researchers published more than a dozen landmark papers on the remarkable abilities of these so-called stem cells. We salute this work, which raises hopes of dazzling medical applications and also forces scientists to reconsider fundamental ideas about how cells grow up, as 1999's Breakthrough of the Year.

Stem cells may one day be used to treat human diseases in all sorts of ways, from repairing damaged nerves to growing new hearts and livers in the laboratory; enthusiasts envision a whole catalog of replacement parts. Despite such promise, many in society object to using stem cells derived from human embryos, a debate that is sure to continue into 2000 and beyond.

But another astonishing development that occurred in 1999 may ease the ethical dilemma. In defiance of decades of accepted wisdom, researchers in 1999 found that stem cells from adults retain the youthful ability to become several different kinds of tissues: Brain cells can become blood cells, and cells from bone marrow can become liver. Scientists are now speeding ahead with work on adult stem cells, hoping to discover whether their promise will rival that of embryonic stem (ES) cells. Thus, 1999 marks a turning point for this young field, as both science and society recognized—and wrestled with—our newfound power to manipulate a cell's destiny.

---

1. Reprinted with permission from *Science Magazine*. Copyright (1999) American Association for the Advancement of Science.

THE DNA STORY

1665: Cells are discovered by British scientist James Hooke.

1859: Charles Darwin publishes his theory of evolution.

1866: Gregor Mendel discovers principle of genetics in studying how plants pass on traits.

1869: Swiss chemist Johann Meischer identifies deoxyribonucleic acid (DNA) in cells, but is unaware of its function.

1944: Team of scientists led by Oswald Avery determine that DNA is the chemical that carries genetic information.

1953: James Watson and Francis Crick determine the double helix structure of DNA.

1966: Identification of the genetic code—A,T, C and G—needed to assemble amino acids into proteins.

1973: First gene cloned at Stanford University; process patented.

1982: First genetically engineered drug approved by FDA: insulin.

1984: First cancer gene identified.

Open Future

Researchers have long been familiar with certain types of stem cells. Back in 1981, they discovered how to culture mouse ES cells, treating them with just the right growth factors to keep them dividing but forever immature. Today such cells are common research tools, used, for example, to produce tens of thousands of knockout mice that lack specific genes. In humans, scientists had managed to identify and culture stem cells from several types of adult tissue, such as bone marrow and brain, but had done little work on stem cells from embryos.

Then last November two company-funded teams announced that they had isolated and cultured human embryonic and fetal stem cells, coaxing them to pause their development before they became committed to any particular fate (*Science*, 6 November 1998, pp. 1014 and 1145). Many scientists worldwide were immediately eager to join the field but were hampered by rules in many countries that restrict public funding for research that destroys a human embryo. Instead, while ES cell research moved ahead in a few privately funded labs, ethical debates on the wisdom of using human embryos in experiments sprang up in all sorts of public forums.

The U.K. imposed a 1-year research moratorium to allow for public discussion, for example, and a French high court advised lifting that country's ban on human embryo research. And a U.S. presidential advisory panel recommended that public funds be available for all types of stem cell research. In proposed rules released this month, the National Institutes of Health is a bit more restrictive, prohibiting researchers from using federal funds to derive cell lines from an embryo but allowing them to work with certain cell lines created using private funds (*Science*, 10 December, p. 2050).

Career Switching

While the public and policy-makers mulled the ethical implications, researchers galvanized by last November's announcements rushed down another avenue of research—on stem cells from adults. Scientists had known for decades that certain kinds of stem cells lurk in adult tissues, for example in bone marrow and skin. But most scientists had assumed that the adult-derived cells have a limited repertoire. Just as years of training usually commit a concert violinist to a career in music, so scientists had assumed that when a young cell takes on an identity and turns various suites of genes on or off, that genetic programming irreversibly commits it to becoming one of just a few cell types.

But this year several startling results have shown that in some cases, those early commitments can be rewritten. In January, Italian and U.S. scientists reported that stem cells taken from the brains of mice could take up residence in the bloodstream and bone marrow and become mature blood cells—a leap roughly equivalent to a music student becoming a successful professional baseball player. That means signals in the immediate environment can in some cases override a cell's history, implying that nature allows developing cells far more freedom than scientists had imagined.

Many scientists initially balked at that idea, but a string of new results seem to back it up. Texas researchers found just a few weeks ago that muscle stem cells could become blood cells, for example. Other scientists reported that stem cells found in rat bone marrow could become liver cells—raising the tantalizing possibility that cells now routinely harvested in bone marrow transplants may have much broader uses. And this fall Pennsylvania scientists reported that mouse marrow cells injected into the brains of newborn mice could develop into brain cells.

As such basic research leaped forward, biomedical scientists sought new ways to use stem cells to help people. For example, researchers this year found that healthy bone marrow stromal cells, which give rise to bone, could strengthen bones and prevent fractures when injected into three children with the bone-weakening disease osteogenesis imperfecta. And in a promising step toward a possible treatment for neural disorders such as multiple sclerosis, researchers in Boston injected neuronal stem cells from healthy mice into the brains of newborn mutant mice that lacked a key protein, part of the protective myelin sheath around neurons. The treated mutant mice produced

1987: Department of Energy publishes recommendations for a 15-year undertaking to map and sequence the human genome.

1990: Human Genome Project begins; goal set of mapping entire genome by 2005.

1992: First physical maps presented for chromosome 21 and chromosome Y.

1993: USDA approves genetically engineered tomato and cow hormone that stimulates milk production.

1995: Physical maps of chromosomes 3, 11, 12, 16, 19 and 22 are published.

1996: Sheep named Dolly becomes first mammal cloned.

1997: Celera Genomics Inc. announces plans to map the human genome.

1999: Ian Wilmut, who cloned Dolly, announces that he's looking for partners in a venture to clone human embryos.

2000: Chromosomes 5, 16 and 19 join 22 as the first to be fully sequenced.

the missing protein and were less prone to a characteristic tremor. Another group injected bone and muscle stem cells into mice that lacked the protein dystrophin—missing in patients with Duchenne's muscular dystrophy—and within 2 weeks the animals produced the missing protein.

Applications of ES cells could be even more dramatic. This year another set of mutant mice, unable to produce myelin, received mouse ES cells in their brains; the cells soon spread through the brain and produced myelin. And Missouri researchers this month reported that mouse ES cells could help restore some movement to the limbs of partially paralyzed rats.

With dramatic results like these coupled with growing public acceptance, the stem cell field is poised for progress. If it lives up to its early promise, it may one day restore vigor to aged and diseased muscles, hearts, and brains—perhaps even allowing humans to combine the wisdom of old age with the potential of youth.

Petri Dish Politics[2]

BY RONALD BAILEY
REASON, DECEMBER 1999

"Death to death," declares Gregory Stock, director of UCLA's Program on Medicine, Technology, and Society, at a conference on life extension. "Aging itself can be considered to be a disease," says Cynthia Kenyon, the biochemist who last year discovered genes that quadrupled the life of the nematode *C. elegans.*

"This is the first time that we can conceive of human immortality," William Haseltine, the hardheaded CEO of Human Genome Sciences Inc., the largest genomics company in the world, tells *The Washington Post.* Francis Fukuyama, the man who famously asserted that "The End of History" had arrived, declares that History is about to begin again, and its motor is biotechnology. "It is no longer clear that there is any upper limit on human life expectancy," writes Fukuyama. That, he argues, changes human nature and thus restarts History.

The biomedical revolution of the next century promises to alter our culture, our politics, and our lives. It promises to extend our life span and to enhance our mental and physical capacities. The closer those promises come to reality, however, the more they incite opposition and, in some cases, horror. And they are becoming more real by the day.

In September, Princeton University neurobiologist Joe Tsien announced that he had boosted the intelligence of mice by inserting extra copies of a gene that produces a type of receptor in brain cells; the receptor enhances long-term memory and learning. The "smart mice" did considerably better than normal mice on a battery of six rodent intelligence tests. The mouse gene Tsien manipulated is 98 percent identical to the one found in humans. In the short term, Tsien's work could lead to drugs that will boost the memory capacities of adult humans. Over the long run, these genes might be introduced into human embryos who, once born, would have an easier time learning and retaining new information. It was the prospect of making smarter people, not just curing Alzheimer's, that made global headlines.

On the horizon are artificial chromosomes containing genes that protect against HIV, diabetes, prostate and breast cancer, and

---

2. Reprinted with permission, from the January 2000 issue of *Reason Magazine.* Copyright © 2000 by the Reason Foundation.

Parkinson's disease, all of which could be introduced into a developing human embryo. When born, the child would have a souped-up immune system. Even more remarkably, artificial chromosomes could be designed with "hooks" or "docking stations," so that new genetic upgrades later could be slotted into the chromosomes and expressed in adults. Artificial chromosomes could also be arranged to replicate only in somatic cells, which form regular tissues, and not in the germ cells involved in reproduction. As a result, genetically enhanced parents would not pass those enhancements on to their children; they could choose new or different enhancements for their children, or have them born without any new genetic technologies.

> *These artificial chromosomes . . . offer exquisite control over which genes will be introduced into an organism and how they will operate.*

Already, a Vancouver company, Chromos Molecular Systems, makes a mammalian artificial chromosome that allows biotechnologists to plug in new genes just as new computer chips can be plugged into a motherboard. These artificial chromosomes, which have been developed for both mice and humans, offer exquisite control over which genes will be introduced into an organism and how they will operate.

Meanwhile, the prospect of substantially extending the human life span is growing, as biomedical researchers investigate promising technologies to diagnose and treat the various ways the body breaks down with age. EntreMed Inc. of Rockville, Maryland, and Cell Genesys Inc. of Foster City, California, are working to deliver a gene-based drug that will cut off a cancer's blood supply and kill it. Human Genome Sciences, also of Rockville, is developing a heart-bypass-in-a-shot using the VEGF-2 gene, which produces a protein that encourages the growth of blood vessels around blocked arteries. In Silicon Valley, Santa Clara-based Affymetrix Inc. has created a "biochip"—a silicon wafer that analyzes thousands of genes in a single test, diagnosing all sorts of diseases. Combined with the full sequence of all human genes, which will be available in a couple of years, the biochip will enable doctors to do a full genetic physical with a simple blood test.

Late last year, Geron Corp. of Menlo Park, California, announced that scientists whose work it had supported had isolated the grail of human cell biology: embryonic stem cells. These remarkable cells are capable of growing into any of the 210 types of cells found in the human body. Suffer a third-degree burn? Grow some skin cells in a petri dish for a skin graft. Heart attack? Replace the damaged tissue with made-to-order heart cells. Broken back? Fix that right up with a skein of new nerve cells.

Repairing broken bodies, extending life, and improving individuals' capabilities may sound like good things. But the promises of biomedicine increasingly attract opposition. A chorus of influential conservative intellectuals is demanding that the new technologies be crushed immediately, and many in Congress are listening. These "luddicons," as one observer has dubbed them, see in biomedicine the latest incarnation of human evil. "In the 20th century, we failed to stifle at birth the totalitarian concepts which created Nazism and Communism though we knew all along that both were morally evil—because decent men and women did not speak out in time," writes the British historian Paul Johnson in an article in the March 6, 1999, issue of *The Spectator*. "Are we going to make the same mistake with this new infant monster [biotechnology] in our midst, still puny as yet but liable, all too soon, to grow gigantic and overwhelm us?"

The most influential conservative bioethicist, Leon Kass of the University of Chicago and the American Enterprise Institute, worries both that our quest for ever-better mental and physical states is too open-ended and, contradictorily, that it is utopian. "'Enhancement' is, of course, a soft euphemism for improvement," he says, "and the idea of improvement necessarily implies a good, a better, and perhaps even best. But if previously unalterable human nature no longer can function as a standard or norm for what is regarded as good or better, how will anyone truly know what constitutes an improvement?"

> *"... the ultimate question is how far we may go in defying nature without undermining our humanity."*— historian **Gertrude Himmelfarb**

Kass argues that even "modest enhancers" who say that they "merely want to improve our capacity to resist and prevent diseases, diminish our propensities for pain and suffering, decrease the likelihood of death" are deceiving themselves and us. Behind these modest goals, he says, actually lies a utopian project to achieve "nothing less than a painless, suffering-free, and, finally, immortal existence."

What particularly disturbs these conservatives is biomedicine's potential to overthrow their notion of human nature—a nature defined by suffering and death. "Contra naturam, the defiance of nature, used to be a sufficient argument for those who were not persuaded by contra deum, provoking the wrath of God," writes historian Gertrude Himmelfarb in *The Wall Street Journal*. "But what does it mean today, when we have defied, even violated, nature in so many ways, for good as well as bad?" She goes on to suggest that cloning, artificial insemination, in vitro fertilization, and even the pill might be "against nature."

Himmelfarb continues, "But the ultimate question is how far we may go in defying nature without undermining our humanity . . . . What does it mean for human beings, who are defined by their mortality, to entertain, even fleetingly, even as a remote possibility, the idea of immortality?" Himmelfarb insists that she doesn't disdain all improvement. "To raise these questions is in no way to reject science and technology or to belittle their achievements," she writes. "It is not contra naturam to invent labor-saving devices and amenities that improve the quality of life for masses of people, or medicines that conquer disease, or contrivances that allow disabled people to live, work and function normally. These enhance humanity; they do not presume to transcend it."

It is hard to see how a genetically enhanced memory, a faster mental processing speed, or a stronger immune system "undermines our humanity." After all, many full-fledged human beings already enjoy these qualities. Nor is it clear why "contrivances" that let disabled people cope with their physical problems are acceptable, while genetic cures to avoid the problems in the first place are not.

Nearly all technologies—agriculture, literacy, electric lighting, anesthesia, the pill, psychoactive drugs, television—affect human nature in the sense that they change the rhythms of human life and widen the range of behavior in which people can engage. We are no longer tribesmen living in family bands of 20, hunting and gathering on the plains of Africa. Surely there have been significant changes in human psychology as a result of the development of civilization. In fact, changing human psychology might be said to be the whole point of civilization; some anthropologists speculate that civilization is a set of social institutions that exist to tame human, especially male, violence.

Himmelfarb and Kass accuse those who favor biomedical progress of seeking immortality, as though that were a self-evident evil. But "immortality" is, in a sense, just a longer life span. Since 1900, life spans worldwide have doubled, and most people think that achievement has been a great moral good. Using genetic techniques to increase human life spans is not any different ethically from using vaccines, organ transplants, or antibiotics to achieve the same goal. Kass and Himmelfarb assert that human beings have been "defined by their mortality." But human beings are perhaps even better defined by their unending quest to overcome disease, disability, and death.

Indeed, all of the things on Himmelfarb's list of acceptable enhancements are "contra naturam." Is it not more natural to tear our meat with our hands rather than with stainless steel forks? Is it not more natural to die by the hundreds of thousands of tuberculosis, smallpox, or ebola? And is it not more natural for the lame, the blind, and old to die beneath the claws and teeth of predators? Himmelfarb does not make it clear how trying to "transcend" the dirty, nasty, brutish, and short lives of our ancestors undermines our humanity. Oh sure, a lower infant mortality rate—down from 300 or 400 deaths per 1,000 live births in the 18th century to only seven per 1,000 today—has deprived us of the chance to contemplate the tragic fleetingness of life and the poignancy of innocent death. But who among us really minds?

In an ironic linguistic twist, the pro-death opponents of substantially extending human life spans have found their greatest allies among the pro-life opponents of abortion. The reason lies in genetic essentialism: the reductionist view of human beings as nothing more than meat puppets dangling from the strands of our DNA. Nowhere is this strange alliance more important, or its philosophical underpinnings more apparent, than in the debate over stem cell research.

In an ironic linguistic twist, the pro-death opponents of substantially extending human life spans have found their greatest allies among the pro-life opponents of abortion.

At the very earliest stages of development, an embryo is an undifferentiated mass of cells, rather than blood cells, neurons, skin cells, muscle cells, etc. These undifferentiated stem cells can develop into any type of tissue. Isolated stem cells could one day be used to grow new heart, nerve, pancreatic, or liver cells that would replace tissues damaged by disease. Such replacement parts could extend human life spans by decades, with significantly improved quality. They are just the sort of ambitious, "unnatural" technologies the luddicons fear.

Currently, biotechnologists investigating stem cells use embryos donated by couples who have had infertility treatments. The embryos are grown in laboratory cultures until they reach the blastocyst stage at four to seven days after fertilization. At that point the embryo consists of about 100 or so cells. A blastocyst is a hollow sphere of cells whose outer layer would develop into the placenta while the inner cell mass grew into a fetus. Once the inner cell mass is extracted from the blastocyst, those stem cells can no longer develop into a complete organism.

Stem cells removed from the blastocyst are grown in a culture on a layer of feeder cells that provide the necessary environment to keep them alive and in an undifferentiated state. Researchers are still trying to learn exactly what molecular signals will cause stem cells to develop into specific tissues. Those signals hold the key to using stem cells to develop replacement tissues which would be part of a universal tissue repair kit.

Once those signals are understood, using stem cells will depend, at least in the near future, on technology originally developed in cloning research. As we know from Dolly the lamb, factors in egg cytoplasm can reset an adult cell nucleus, giving it the ability to grow into an embryo as a source for stem cells. Using cloning technology, doctors might one day take the nucleus of one of your skin cells, put it in a human egg from which the nucleus has been removed, and allow that cell to divide to the blastocyst stage. They would then take out the stem cells from its inner cell mass and dope them with the appropriate hormones and proteins to turn the stem cells into, say, heart tissue, which could then be used to repair your ailing heart. Using your own cells in this way would mean that your immune system wouldn't reject the newly engrafted tissues, since the tissues would be a perfect match.

This research obviously promises to significantly advance human health and longevity. And just as obviously, stem cell research is completely entangled with the politics of abortion. It involves the use of embryonic tissues and, eventually, the creation of fertilized eggs that abortion opponents consider full-fledged human beings. To abortion opponents, a blastocyst used to duplicate your heart tissue isn't an extension of your tissue. It's another human being—the equivalent of your identical twin. As Judie Brown, president of the American Life League, told the *Los Angeles Times* about research on embryonic cells, "It doesn't matter if it's done in the womb or a petri dish, it's still killing."

> *This research obviously promises to significantly advance human health and longevity.*

After Geron scientists announced in November 1998 that they'd isolated human embryonic stem cells from donated embryos and aborted fetuses, President Clinton asked the National Bioethics Advisory Commission to look into any ethical issues associated with stem cells. In January, the U.S. Department of Health and Human Services ruled that the National Institutes of Health could fund research using already derived embryonic stem cells. This ruling provoked 70 anti-abortion House members, including Majority Leader Richard Armey (R-Tex.), Majority Whip Tom DeLay (R-Tex.), and Republican Conference Chairman J.C. Watts (R-Okla.) to sign a letter of protest to the president, declaring that the HHS ruling violated the congressional ban on funding research on human embryos. The congressional ban, adopted in 1996, outlaws the use of federal funds for the creation of human embryos for research in which they are "destroyed, discarded or knowingly subjected to risk of injury or death."

In January, however, HHS General Counsel Harriet Rabb artfully concluded that Geron's embryonic stem cells "are not a human embryo within the statutory definition." She based her decision on the fact that the cells "do not have the capacity to develop into a human being, even if transferred to the uterus." Consequently, destroying them in the course of research would not constitute the destruction of an embryo.

NIH scientists, whose work depends on federal funding, are eager for the ban to be lifted. NIH Director Harold Varmus correctly claims that federal funding also brings federal oversight, which he argues will protect the public interest. Of course, Varmus and other researchers curiously overlook the point that it was precisely federal oversight that led to the ban on federal support of this important research in the first place.

As it became clearer that the National Bioethics Advisory Commission was going to recommend that some stem cell research be federally funded, opponents turned up the heat. In July, Sen. Sam Brownback (R-Kan.) sponsored a Capitol Hill press conference featuring a group of bioethicists, religious activists, and physicians who oppose human embryonic stem cell research. "Human embryos are not mere biological tissues or clusters of cells; they are the tiniest of human beings," asserts the group's July 1 press release.

At the press conference, Edmund Pellegrino, a bioethicist at Georgetown University's Kennedy Bioethics Center, took aim at even private research efforts like those sponsored by Geron. He urged that a congressional ban "should be extended permanently to include privately supported as well as federally supported research involving the production and destruction of living human embryos." Although the current debate centers on federal funding, the real issue is whether the research should be done at all.

People who oppose stem cell research on the ground that any cell that can become a human being already is a human being are essentially arguing that every cell in your body is another person.

According to an NIH spokesperson, the NIH's draft guidelines for embryonic stem cell research are likely to be issued before the end of the year, and Congress will take up the subject in February. Sens. Arlen Spector (R-Pa.) and Tom Harkin (D-Iowa) are the two leading proponents of human stem cell research. Opponents include DeLay, Brownback, and Rep. Henry Hyde (R-Ill.).

The opponents argue that biotechnologists should concentrate on isolating and using stem cells known to exist in adults instead. Such adult stem cells are the precursor cells that renew tissues like skin and the linings of the intestines, and they likely could be used to regenerate these tissues. But many researchers believe that adult stem cells won't be as protean as embryonic cells—that they won't be able to turn into as many different types of cells.

One day it may be possible to take any adult stem cell back to the embryonic, and hence protean, stage. But the research to figure out how to do that depends on work with embryonic cells and the resulting cells, of course, would themselves be embryonic. People who oppose stem cell research on the ground that any cell that can become a human being already is a human being are essentially arguing that every cell in your body is another person.

"What happens when a skin cell turns into a totipotent stem cell [a cell capable of developing into a complete organism] is that a few of its genetic switches are turned on and others turned off," writes University of Melbourne bioethicist Julian Savulescu in the April 1999 issue of the *Journal of Medical Ethics*. "To say it doesn't have the potential to be a human being until its nucleus is placed in the egg cytoplasm is like saying my car does not have the potential to get me from Melbourne to Sydney unless the key is turned in the ignition." Since nearly every cell in the human body contains the complete genetic code of an individual, it is logically possible using biotech to turn every one of a person's cells into a complete new human being. If one doesn't turn on the ignition of a car (or one doesn't strip the suppressor proteins from a nucleus and put the cell into a womb), then the car won't go (or the skin cell won't grow into a human being). In other words, simply starting a human egg on a particular path, either through fertilization or cloning, is a necessary condition for developing a human being, but it isn't sufficient. A range of other conditions must also be present.

> *"Cloning forces us to abandon the old arguments supporting special treatment for fertilised eggs."*—
> bioethicist **Julian Savulescu**

"I cannot see any intrinsic morally significant difference between a mature skin cell, the totipotent stem cell derived from it, and a fertilised egg," writes Savulescu. "They are all cells which could give rise to a person if certain conditions obtained." Those conditions include the availability of a suitable environment like a woman's womb. A petri dish is not enough.

"If all our cells could be persons, then we cannot appeal to the fact that an embryo could be a person to justify the special treatment we give it," concludes Savulescu. "Cloning forces us to abandon the old arguments supporting special treatment for fertilised eggs."

The DNA content of a skin cell, a stem cell, and a fertilized egg are exactly the same. The difference between what they are and what they could become is the environment in which their DNA is found. Thus, Savulescu argues, the mere existence of human DNA in a cell cannot be the source of a relevant moral difference. The differences among these cells are a result of how the genes in each are expressed, and that expression depends largely on which proteins suppress which genes. Does moral relevance really depend on the presence of the appropriate proteins in a cell? Trying to base moral distinctions on this level of biochemistry seems a bit quixotic.

So, asks Savulescu, is it immoral for you to take one of your skin cells, put it into an enucleated egg, and begin to grow it in a petri dish with the intention of making new brain cells to cure your Parkinson's disease? The cell was your tissue, with your genes. The transformed cell would not exist except for your intention—it would simply have flaked off and gone down the drain. "It's important to remember that essentially every cell in our body has a full complement of genes and in that sense is potentially totipotent," Varmus, the NIH director, reminded the National Bioethics Advisory Commission. That a cell contains a complete set of human chromosomes, yours, surely does not make that cell the moral equivalent of a baby. But as Savulescu and Varmus point out, if one is committed to the sort of genetic essentialism relied on by many opponents of cloning and embryonic stem cell research, then one is also logically committed to maintaining that the only difference between your skin cell and your twin is which proteins decorate their DNA strands.

Once it is possible to make stem cells without eggs, perhaps the moral intuition of many people will shift.

The next step in stem cell research will occur when biotechnologists learn how to strip off the suppressing proteins from a mature cell's genes and transform it directly into a stem cell without having to use enucleated human eggs. That advance will take human eggs out of the discussion. Once it is possible to make stem cells without eggs, perhaps the moral intuition of many people will shift.

"It may eventually become possible to take a cell from any one of our organs and to expose it to the right set of environmental stimuli and to encourage that cell to return to a more primitive stage in the hierarchy of stem cells," explains Varmus. "Under those conditions, one might in fact generate the cell with as great a potential as a pluripotent cell [capable of becoming many different, but not all, types of tissues] from a very mature cell. One might even in fact imagine generating a cell that is totipotent in that manner." (Again, a totipotent cell is one that could develop into a complete organism if put in the right circumstances.)

Stem cells produced this way would be identical to the human embryonic stem cells that currently must be harvested from embryos. A cell whose suppressor proteins have been stripped off could become a nerve stem cell, a liver stem cell, or a baby—

depending on the intentions of the patients and doctors. Researchers are experimenting right now to see if new embryonic stem cells could be formed by introducing the nucleus of an adult cell into an already existing enucleated embryonic stem cell, thus bypassing the need to use human eggs.

One final consideration is that those committed to claims that individual human beings are defined by their DNA must take into account the fact that up until the eight-cell stage any one of an embryo's cells could become a separate embryo and, under the right circumstances, develop into a

> *What makes us distinct and unique is not our genes but our brains and the minds they contain.*

baby. So until that point are there several persons, or one, in a fertilized human egg? It is now possible after a fertilized egg has first divided into two cells to take one cell and use it to test for genetic diseases. The tested cell could have developed into a baby if placed in a woman's womb. Has the genetic test killed a twin?

Stem cell research opponents might respond that these arguments are just splitting hairs. But there are quite a lot of biochemical hairs to split. And just how you split them determines how you regard the moral status of all types of cells.

Does human uniqueness really reside in our genes? Try this thought experiment. Imagine that transplant surgery has so improved that it is possible to remove your brain and place it safely into another body. So after your brain transplant, is your original body "you?" Or are "you" residing now in a different body? A new body could certainly change "you" in certain ways, since your senses and biochemistry would be different. But humans already affect the operation of their brains by giving themselves different drugs. When people take therapeutic drugs, say Prozac, L-dopamine, or even steroids, we do not believe they become different people.

"The human mind, of course, is a dynamic entity, but genes are static," explains Princeton biologist Lee Silver. A person's genes provide the instructions for building his or her brain, and the mind which comes out of that brain can respond to the environment. Unlike genes, a mind can change. "The human mind is much more than the genes that brought it into existence," concludes Silver.

What makes us distinct and unique is not our genes but our brains and the minds they contain. Persons generally have brains that are capable of supporting enough mental activity to give rise to a mind. As one of my old philosophy professors once put it, "I have never seen a mind that was not located in fairly close proximity to a brain."

The point that brains, not genes, are the source of our unique-ness is further underscored by the fact that no one argues that natural clones, otherwise called identical twins, are the same per-son, even though they share an identical set of genes. They have two different brains and experience the world from two different points of view. Human brains—malleable, fluid, flexible, changing—not static genes, are the real essence of what defines us as peo-ple. We are not mere meat

> *Clearly we must be on guard against any attempts to harness this new technology to government-mandated ends.*

puppets at the mercy of our genes. In fact, with biotech it might better be said that our genes are now at the mercy of our minds.

Leon Kass is disheartened by this prospect. "We triumph over nature's unpredictabilities only to subject ourselves, tragically, to the still greater unpredictability of our capricious wills and our fickle opinions," writes Kass in the September issue of *Commentary*. In other words, he is against human freedom because he doesn't think we can handle it. Ultimately, Kass wants to preserve the "freedom" of some portion of humanity to be miserable, sick, and unhappy. But if they were truly free, would people choose to suffer or to subject their children to such suffering? Not likely.

Kass does have a point, however, when he writes in *Commentary*, "Even people who might otherwise welcome the growth of genetic knowledge and technology are worried about the coming power of geneticists, genetic engineers, and, in particular, govern-mental authorities armed with genetic technology."

There is a threat of government control. Some intellectuals are already succumbing to the temptation of government-supported and mandated eugenics, lest the benefits of genetic engineering be spread unequally. "Laissez-faire eugenics will emerge from the free choices of millions of parents," warns *Time* magazine colum-nist Robert Wright. He then concludes, "The only way to avoid Huxleyesque social stratification may be for government to get into the eugenics business."

Clearly we must be on guard against any attempts to harness this new technology to government-mandated ends. But a Brave New World of government eugenics is not an inevitable conse-quence of biomedical progress. It depends instead on whether we leave individuals free to make decisions about their biological futures or whether, in the name of equality or of control, we give that power to centralized bureaucracies. Huxley's world had no "laissez-faire eugenics" emerging from free choice; *Brave New World* is about a centrally planned society.

A biological future without a plan is exactly what scares critics on both the right and the left. "Though well-equipped [through biotech], we know not who we are or where we are going," Kass fearfully writes. If we know not who we are, surely advances in biotech

---

Like all technologies, biotech could be abused, but using it is not . . . the same as abusing it.

---

are helping us to understand more completely who we are. As for where we are going, the fact that we don't know is why we go. Over the horizon of human discovery Kass sees a territory marked, like the maps of yore, "Here be monsters." To avoid the supposed monsters, Kass wants humanity to stay quietly at home with its old conceptions, technologies, traditions, and limited hopes.

If we use biotech to help future generations to become healthier, smarter, and perhaps even happier, have we "imposed" our wills on them? Will we have deprived them of the ability to flourish as full human beings? To answer yes to these questions is to adopt Rousseau's view of humanity as a race of happy savages who have been degraded by civilization. The fact is that previous generations have "imposed" all sorts of technologies and institutions on us. Thank goodness, because by any reasonable measure we are far freer than our ancestors. Our range of choices in work, spouses, communities, medical treatments, transportation—the list is endless—are incomparably vaster than theirs. Like earlier technologies, biotech will liberate future generations from today's limitations and offer them a much wider scope of freedom. This is the gift we will give them. Like all technologies, biotech could be abused, but using it is not, as Kass and Paul Johnson would have us believe, the same as abusing it.

Scientific facts will not resolve these issues. On the one hand, people who see human genes as the defining essence of humanity will object to stem cell research and a good deal else in the coming biotech revolution. On the other hand, people who see human beings as defined essentially by their minds will have fewer moral objections.

At a hearing earlier this year, Edward Furton, who works at the National Catholic Bioethics Center, asked the National Bioethics Advisory Commission to "please remember in your deliberations that millions of your fellow citizens hold that the human embryo is a human life worthy of the protection of the law." He added, "As a result of the tainted origin, many Americans who have deeply held moral objections to embryo destruction may choose not to receive any benefits from this new research."

No one is suggesting that people should be forced to use medicines that they find morally objectionable. Perhaps some day different treatment regimens will be available to accommodate the different values and beliefs held by patients. One can imagine one medicine for Christian Scientists (minimal recourse to antibiotics, etc.), another for Jehovah's Witnesses (no use of blood products or blood transfusions), yet another for Roman Catholics (no use of treatments derived from human embryonic stem cells), and one for those who wish to take the fullest advantage of all biomedical discoveries.

In a sense, the battle over the future of biotech—and, if Fukuyama is correct, the future of humanity—is between those who fear what humans, having eaten of the Tree of Knowledge of Good and Evil, might do with biotech and those who think that it is high time that we also eat of the Tree of Life.

Immortality and Its Discontents[3]

BY SHERWIN B. NULAND
WALL STREET JOURNAL, JULY 2, 1999

One way or another, scientists will see to it that we live longer. Having gradually added more than 30 years to our average life expectancy since 1900, they are now beginning to predict astonishing accelerations in their success. The most optimistic of them have us reaching ages of 150 and more—much more, say some—by the end of the 21st century.

By far the greatest progress in life-lengthening until now has been made by conducting campaigns against lethal factors invading our bodies from the outside. Improvements in water supply, personal cleanliness, living conditions and immunizations have done far more for longevity than have even the most dramatic of clinical advances, with the exception of antibiotics. Most of these efforts had their greatest impact in the first half of the century.

In more recent decades, increased awareness of other kinds of environmental killers has resulted in further significant gains. By stressing the dangers of cigarette smoking, solar radiation, alcohol consumption, environmental toxins and sedentary habits, authorities have successfully stressed, at least in industrialized nations, the role that each of us can play in improving our own health.

While all of this was going on, bedside medicine was undergoing changes of enormous magnitude. With the advent of the research and clinical applications of molecular biology beginning in the 1960s, truly wondrous progress has been made in the treatment of the sick. Because such therapies help only one person at a time, the consequent drop in mortality statistics is significantly smaller than what has been accomplished by the preventive measures, but striking nevertheless. When improved patient care is combined with other recent changes, the results can be impressive. Witness how developments in rapid-response medical transport, intensive-care units, bioengineering and awareness of the necessity for lifestyle changes have intermixed with rapid advances in more classical medical and surgical treatment to decrease the mortality from heart attacks by more that 30% in the past three decades.

But the visionary biomedical scientists have lately begun to insist that everything they have done for us till now is just chickenfeed.

In genetic engineering, we are told, lies the promise of the brave new world in which there will be so many people of very advanced age that the U.S. Postal Service will not be able to print enough stamps to keep up with the needs of the AARP. This will not come about by the small increments of yesteryear, but through new approaches so profound in their implications that only the most immediate of them can yet be imagined.

The seers of science forecast that genes affecting senescence and death will be switched off and on at will, or snipped out and replaced by more compliant bits of DNA. But ingenuity will not end there, they claim. Certain general determinants—or at least indices—of the aging process will also come under control. The most heralded is the telomere, best described as a kind of cap on the end of each chromosome. As a cell ages, its telomeres get smaller, but the process can be stopped or reversed by a recently identified enzyme called telomerase. Experiments with telomerase in the laboratory have resulted in significant increases in the number of times a cell can reproduce before dying. It is thought that the added longevity of a cell may eventually translate into the added longevity of a whole person.

> *It is thought that the added longevity of a cell may eventually translate into the added longevity of a whole person.*

And then, of course, there are all kinds of breathtaking predictions associated with the application of the coming embryonic stem cell research. Everything from the replacement and repair of small segments of sick tissues to the creation of entire organs has been projected to be possible in the near future. It is difficult to believe that having been successful in replacing what is sick, the scientists will not succumb to the temptation to replace what is merely getting old. We can bet that the human predisposition to seek immortality will create intense pressures on the scientific endeavor. Our shared propensity for narcissism will add its contributions to the clamor.

But before the big push for the new longevity begins, there are already some serious problems on the horizon. Even should none of the anticipated technologies pan out, current trends indicate that the population beyond age 65 in developed countries will rise to 25% by 2050, from 14% today. Just suppose that genetic engineering and the other methods do prove to be successful. What will the U.S. be like in 2100 if the predictions of the molecular manipulators come true, and even as much as a fifth of its present citizenry is still alive? Let others direct themselves to the social, economic and demographic consequences of such a situation. What I am concerned about is less the absolute number of the aged population than the health status of its individual members.

It is unimaginable that even the manipulation of DNA will result in a world in which centenarians, their parents and their grandparents are free of the debilities of advancing age. Can dementia, arthritis, lessened immunity, susceptibility to malignancy, vision and hearing problems and depression be somehow prevented in these people who live longer? I think not. To accomplish a miracle of that sort would require a technology capable of rejuvenating all tissues at once, and not even the most starry-eyed of the futurists is ready to go that far—and it can be predicted that they never will.

Without a doubt, quite the opposite will be the case. We will be faced not only with a large number of debilitated people, but very likely with new medical problems at whose nature we can now only guess. These are as-yet unforeseeable sicknesses that will not have previously manifested themselves, because one must live very long for them to appear. The very fact of aging carries a genetic predisposition to certain diseases that are less likely to be encountered in younger individuals. Just think of how many pathologies have markedly increased in frequency already, as a direct consequence of a population living long enough to develop them. Plenty of cancers fall into this category. It will never be possible, and in any event not desirable, to change the entire biology of aging.

We will be faced not only with a large number of debilitated people, but very likely with new medical problems at whose nature we can now only guess.

I refer to the obvious fact that changing the biology of aging cannot be accomplished without playing fast and loose with more than just a little of the vast panorama of the human genome. Altering some individual snippets of DNA will not do the trick. Quite clearly, there are limiting factors on just how much tinkering the 3.5 billion year-old evolution of our heredity will tolerate, and I suspect that it is not much. Scientists can go just so far toward upsetting the delicate and interdependent balance that is our biology, and no further. When they have gone to their ultimate point of daring to make us live longer, the world will not be filled with 150-year-olds who function as though they are 45. Instead, it will be heavy with the hoary.

If we really want to make our lives better—and by this I mean the entire length of an increasingly long life—perhaps we are looking in the wrong direction. The single most important factor determining whether a person can live independently in old age is not any named disease, but physical frailty. As one gets older, progressive loss of muscle strength increasingly limits one's ability to avoid sickness. What is most astonishing about this now well-accepted observation is that it has only recently been made. A program including aerobic and weight resistance exercises for the oldest old has repeatedly been shown to improve the health and functioning of the heart and lungs, sharpen intellectual abilities, build bone density, decrease osteoporosis, enhance the immune system, and even lower the death rate from some cancers—not to mention its salubrious effect on mood and self-image. This is just one example of how we might best direct our efforts to prolong life, and improve its quality.

In the late 1970s, when physicians believed with seeming good reason that infectious disease had been all but conquered, the talk was of turning the main thrust of future research toward the goal of preventing and treating the processes of degeneration that can begin at any age, but are most likely to hit the elderly. As our population becomes ever older, even before major efforts to toy with our DNA get under way, we would do well to look back on that relatively modest goal, and return it to the pre-eminent place it once occupied on the A-list of our research priorities. We have become so enthralled with the accomplishments of the basic scientists who tell us that they can make us live longer that we have sadly neglected the equally impressive accomplishments of those who have continued to strive so that we may live better.

A Thriving Pioneer of Gene Tests[4]

By Marlene Cimons
Los Angeles Times, April 17, 2000

". . . A great day for the world. A great day for medicine. Gene therapy has been approved. Our daughter is the first patient in the world to receive it."

Those were the words that a tired but exhilarated Van DeSilva penned into her personal journal on the evening of Sept. 14, 1990. Her 4-year-old daughter, suffering from an extremely rare and fatal disease that cripples the immune system, had just made medical history.

Ten days earlier, at the National Institutes of Health in Bethesda, Md., Ashanthi DeSilva sat quietly watching the children's adventure fantasy movie "Willow" as researchers repeatedly drew blood from her tiny arm. They extracted white blood cells from the samples, then transferred those into tissue culture in the lab. As the cells grew from tens of millions to billions, scientists added a corrected cell that—it was hoped—would do the work that her own flawed genes could not.

On the day that her mother wrote those uplifting words, Ashanthi was injected with the cells containing the healthy genetic message. The daring experiment in gene therapy had seemingly gone well and scientists celebrated their arrival on the cutting edge of a new medical frontier.

Today, Ashi, as she is known, is a thriving teenager in Ohio, a 13-year-old who gets all A's and B's, plays the piano and basketball and rarely gets a cold.

She is not cured: Ashanthi still must take a drug that helps keep her disorder under control. But her body produces some cells with the new gene and is faring better than scientists had dared imagine 10 years ago.

Gene therapy, however, is under siege. Federal agencies have begun to crack down on experiments, after the unexpected death of a man who was not fatally ill when his gene therapy began. Scientists and government officials have expressed skepticism about cases like Ashi's, arguing that anecdotes do not constitute proof. And the public has grown weary of New Age medicine, worried that

4. Article by Marlene Cimons from the *Los Angeles Times* April 17, 2000. Copyright © *The Los Angeles Times.* Reprinted with permission.

the early promise of high-tech science has given way to unacceptable risks.

Now, scientists again look to Ashi to validate their case.

Sometime this summer, doctors again will take blood samples from her, again grow them in the lab and again genetically change the cells and return them to her body.

But this time the corrected cells will be stem cells—the purest

---

"This field was supposed to be magic. And people think that, if you don't reach a high level of expectation, you are a failure."—Dr. R.
Michael Blaese

---

kind. This time, researchers will transplant them into her bone marrow, hoping that the marrow, from then on, will stop producing the genetically defective cells and only produce those that are genetically perfect.

And this time, they hope that they finally will have their proof.

Dr. R. Michael Blaese, the former NIH researcher who treated Ashi in 1990 and is designing the new, more advanced therapy she will receive soon, is baffled by the current furor over gene therapy. He believes that, without the therapy she received, Ashi likely would have died in childhood from her disorder, which left her vulnerable to one infection after another.

While her body is still making cells with the bad genes, many of the good genes inserted years ago are still circulating in her system, helping to keep her healthy, he said. He thinks that the public is holding gene therapy research to an unrealistically high standard.

"This field was supposed to be magic. And people think that, if you don't reach a high level of expectation, you are a failure," he said.

Success Measured in Increments

"The reality is that success often comes in increments. Something may be of value, even if it's not a cure. To say that gene therapy has never helped anyone is simply false."

The original rationale behind gene therapy was to treat genetic diseases by replacing a non-functioning or defective gene, which was the case with Ashi. While gene therapy experiments still focus on genetic disorders, the field has broadened considerably in recent years to look at gene therapy approaches to a range of other diseases, among them cancer, heart disease and AIDS.

Since 1990, 4,000 patients have participated in 378 gene therapy trials, of which 343 are still underway. There are 32 gene therapy trials ongoing in California, according to federal health officials. Some of the sites in Southern California are UCLA, City of Hope National Medical Center in Duarte, Cedars-Sinai Medical Center, Scripps Clinic in La Jolla, USC, Children's Hospital Los Angeles and Sidney Kimmel Cancer Center in San Diego.

Gene therapy research became controversial after 18-year-old Jesse Gelsinger of Arizona died in September after receiving experimental gene therapy for a liver disorder at the University of Pennsylvania's Institute of Gene Therapy.

Since his death, federal health officials determined that the institute violated federal regulations in the conduct of the study. They halted all gene therapy there and began to tighten oversight on all researchers doing gene therapy experiments.

For Ashanthi's coming experiment, her doctors—as they are required to do—sought and received permission from the Food and Drug Administration months ago. And they also won the blessings of an NIH advisory panel that oversees gene therapy research. Approval by the Recombinant DNA Advisory Panel is no longer required as it was in the early days of gene therapy. But the panel has the right to request a review, and it did, likely because "of the current climate," Blaese says.

Federal health officials acknowledge that the field has produced some encouraging results, recent among them reports on treating hemophilia and in bypassing clogged coronary and leg arteries.

But, overall, until there is definitive scientific proof, they tend to dismiss many such cases as anecdotal.

"Hints of Success"

"We do have some hints of success, but those are anecdotal studies," said Lana Skirboll, NIH's director of science policy, who is one of the officials in charge of her agency's investigation of gene therapy practices.

"The scientific community is hesitant to talk about successes without peer review and duplication," she added.

Usually, however, science requires carefully designed clinical trials with "controls" for comparison purposes and large numbers of volunteers so that statistically significant differences can be documented.

But highly experimental gene therapy has been tested for the most part only on small numbers of very ill people, who often have died from disease—making it difficult to prove success.

In 1993, for example, Blaese conducted a gene therapy trial on 15 patients with glioblastoma, a deadly brain cancer. There were no controls. All of the patients were very ill and had failed conventional therapy.

All 15 received experimental gene therapy. They were given a gene designed to change the metabolism of the cancer cells, making them resemble a herpes virus, so they would be vulnerable to

---

. . . highly experimental gene therapy has been tested . . . only on small numbers of very ill people, who often have died from disease— making it difficult to prove success.

---

the antiviral drug acyclovir. The hope was that cells treated with the gene would be killed by the drug.

Fourteen died, victims of their brain cancers. One man, however, is alive and cancer-free seven years later. Gene therapy was a last resort for a dying man—he had had surgery, chemotherapy and radiation, all of which had failed. The cancer had returned in three separate sites.

Would he have died without gene therapy? Blaese thinks so but he cannot prove it.

"He had a bad brain tumor and it came back. He had gene therapy and it went away," said Blaese, who now conducts genetic research for a private company, Kimeragen Inc. in Newtown, Pa., but still serves as an NIH consultant.

"It's very difficult to say with certainty that the gene therapy cured him. He might have been luckier than hell to respond when others didn't. It wasn't a controlled clinical trial, so the scientific community dismissed it as unproven," Blaese said.

"One anecdotal case of a 'cure' gets no attention," he added. "But those same people who dismiss it will jump on a single anecdotal case of a terrible side effect, or death, and paint the entire field with a broad brush. And that's very frustrating."

Ashanthi's parents, originally from Sri Lanka and now living in North Olmsted, a suburb of Cleveland, are saddened that Gelsinger's death has thrown pessimism over a field they believed saved their daughter.

"I feel very sorry for Jesse's family," said Van DeSilva, a homemaker whose husband, Raj, works as a chemical engineer. "A child is dead and that is so very sad. My heart is broken for his parents. I'm sure his mom and dad feel a lot of pain.

"But what am I going to say? I cannot say anything bad about gene therapy. It worked for us," she added. "In my heart, my little girl was cured that day—and the part of my heart that belonged to her was healed too."

Ashi became ill soon after birth. She began suffering persistent bouts of diarrhea and vomiting, rashes that covered her body and infections, one after another.

It took two years to figure out what was wrong with her. She was diagnosed with adenosine deaminase deficiency, an extremely rare genetic disorder that cripples the immune system. Only a few dozen children in the world have it.

It involves a defect in the gene responsible for producing an enzyme critical to the functioning of the immune system—adenosine deaminase.

The enzyme is responsible for clearing the body of a toxin that is produced in the normal course of cell metabolism. If the body cannot rid itself of the toxin, deoxy-adenosine, it poisons the cells of the immune system. Without this enzyme, the body cannot fight infections. Ashi was developing high levels of the toxin, which was causing her immune system to fail.

Doctors at Rainbow Babies Hospital in Cleveland knew that NIH researchers—Blaese and Dr. W. French Anderson, now at USC, and Dr. Ken Culver—were exploring the possibility of using gene therapy to treat the disorder. But the idea had never before been tested.

Listening to Blaese and the others describe the potential and the risks, Ashi's mother kept thinking: "I want this treatment for my daughter."

She said that she was never frightened. Instead, "I was thrilled. I was grateful. I was happy for my little girl."

At age 2, Ashi had begun taking a drug, PEG-ADA, that helped replace the missing enzyme and rid her body of some of the deadly toxin. Still, immune system tests on her showed that "things were not going so well," Blaese said.

"The enzyme helped but I was very worried for Ashi that she would ultimately succumb to her disease, even though she was getting the drug," Blaese says.

Others had failed to survive on the drug, which must be taken for life. And it is very expensive—as much as $400,000 a year, depending on the dose.

On Aug. 1, 1990, the three scientists received approval for their groundbreaking experiment from the Recombinant DNA Advisory Committee. In the drama of making its decision, committee members described the moment as "truly a historic occasion."

On Sept. 4, 1990, the DeSilvas brought Ashi to NIH, where researchers took blood from her arm four separate times. She was very stoic, rarely crying during the repeated blood drawings. From her blood, cells were extracted that later would be put back into her body containing the new genes.

"There are a lot of things that go through your mind when you prepare to do something you've never done before," Blaese remembered. "You agonize over whether you've done everything. . . . What scares you is if there's something you haven't thought about."

Nothing went wrong. It was routine.

In fact, "she seemed like a happier kid the next day," her mother recalled. "She was hugging us and kissing us. I remember telling the doctors, 'I think you gave her the 'happy' gene.'"

Ashi stayed at NIH three days so she could be closely watched, then was sent home. Three weeks later she came back and the process was repeated.

Over the next two years, she received 10 more treatments—monthly at first, then gradually stretched out over more months. Her last session was in August 1992.

Nearly eight years after she received her final treatment, Ashi still has a high proportion of the new, corrected cells circulating in her body, doing what they are supposed to do.

"We looked at the level of enzyme produced by her T-cells, and before she had the treatment, it was zero," Blaese said. "After each treatment, the number of cells producing the enzyme increased in number. In August 1992, more than 50% of her circulating cells carried the gene.

Ashi still receives a weekly injection of the drug, administered by her mother. But the dose, which is typically given by body weight, is the same as it was 10 years ago—meaning that she needs far less of it today than she otherwise would.

Some still question whether she is doing well because of the experiment or because of the drug or because of both. No one is willing to take the risk of stopping the drug.

Dr. Donald Kohn, director of gene therapy at Children's Hospital Los Angeles, in 1993 tried a procedure similar to the one that Ashi soon will receive on three newborns with the same immune disorder as Ashanthi. His team transplanted gene-corrected stem cells obtained from the babies' umbilical cord blood when they were only 4 days old.

That experiment proved safe but disappointing. Only a few new corrected blood cells were produced. The three babies he worked with—approaching age 7—are still taking the drug PEG-ADA.

But he agrees with Blaese that improved techniques could make a big difference this time, adding that all of these attempts represent "very important, positive steps."

There are better methods for growing cells, and a more active retrovirus to use to introduce the new cells into the body, raising the odds that it will work.

And, if so, Ashi's body will stop making cells with the defective gene and only make cells with the right gene—in other words, a cure for her—and a boon for gene therapy.

IV.

Send in the Clones

Editor's Introduction

The 1978 film *The Boys from Brazil* painted one of the most terrifying scenarios anyone in the world could imagine. What if Nazi scientists who escaped Germany after World War II could clone Hitler? Indeed, at the time the movie was released, modern science and genetic research were progressing at such a rate as to make the cloning of human beings seem possible—eventually. Even while audiences shivered at the thought of reproducing an exact replica of the man many considered the most evil human being to have ever lived, most people took comfort in the notion that they would probably never see the day when it would be possible. It was too bizarre, too science-fictional. Then, not even 20 years later, Dolly was created. Practically overnight, this young sheep, cloned by Dr. Ian Wilmut of the Roslin Institute in Edinburgh, Scotland, reignited old debates, rekindled old fears, and stimulated untold imaginations. The unthinkable was suddenly before us: If one species of mammal could be cloned, what next? Who next? What *is* a clone, anyway? How much of a human being can *really* be copied? Religious leaders wondered, would the clone have a soul? If so, who's soul would it be—its own, or that of its original? As thoughts began spinning, scientists and politicians began a dialogue designed to slow the progress of cloning technology and introduce some sanity into the emotional and moral chaos that Dolly's creation had brought into the world. The articles in Section IV present a range of responses to the idea of cloning and the applications of its technology.

This section's first article is Gina Kolata's report on Dolly's creation days after it was announced in February 1997. Her *New York Times* article entitled "With Cloning of Sheep, Ethical Ground Shifts" provides an excellent introduction to the major issues and questions raised by Dr. Wilmut's breakthrough and is therefore an appropriate place to begin exploring this topic. A diverse group of opinions are presented, from one professor's hailing of the "wonderful benefits" for humankind that will come from this innovation, to another's accusation that Wilmut and other researchers are "not thinking about the implications of their work." As Kolata reports, while Dolly's creation brings us one step closer to genetic engineering—a fearful prospect for many— it could also aide humankind in the fight against disease and infertility.

In the article which follows, Marjorie Miller of the *Los Angeles Times* revisits Dr. Wilmut three years after Dolly's birth to discuss his views on the technology he did so much to advance. Her interview with Wilmut, entitled "For Cloning Pioneer, Biotechnology Holds Promise of Medical Gains," discusses the way his newfound fame has affected his life and work, the misconceptions associated with cloning, and what he sees as its objectives. While he is entirely against human cloning, Wilmut continues to tout the benefits of cloning live-stock and remains positive about the field of biotechnology.

Charles Krauthammer's essay "Of Headless Mice . . . and Men," first published in *Time*, considers one of the more troublesome applications of cloning: the growing of headless people for organ transplants. The breeding of human clones for this purpose is a deed that even Dr. Wilmut cannot countenance, yet, Krathammer claims, many scientists discuss doing this very thing. Krauthammer argues vigorously against such a practice, which he believes will have a domino effect of inevitably leading to wide- scale human cloning.

Although some individuals take comfort in the belief that human cloning is merely a theoretical possibility and that all nations will abide by the human cloning moratorium called for by the governments of the United States and Europe, there is resistence, as expressed by Stephen G. Post in his article for *America* entitled "The Judeo-Christian Case Against Human Cloning." Post considers seven moral issues related to this field of biotechnology, among them the manner in which it would make males of the species unnecessary to the process of procreation, the danger of one generation exerting what he calls "overcontrol" or "overpower" upon the next, and his central issue, respect for "Nature and Nature's God." Throughout, Post supports his position with references from the Bible, as well as quotations from scholars from such diverse fields as constitutional law, psychology, theology, literature, fertility science, and bioethics. Though writing from a Judeo-Christian perspective, Post echoes some of the most common objections found in the cloning debate.

With Cloning of a Sheep, the Ethical Ground Shifts[1]

BY GINA KOLATA
NEW YORK TIMES, FEBRUARY 24, 1997

When a scientist whose goal is to turn animals into drug factories announced on Saturday in Britain that his team had cloned a sheep, the last practical barrier in reproductive technology was breached, experts say, and with a speed that few if any scientists anticipated.

Now these experts say the public must come to grips with issues as grand as the possibility of making carbon copies of humans and as mundane, but important, as what will happen to the genetic diversity of livestock if breeders start to clone animals.

For starters, quipped Dr. Ursula Goodenough, a cell biologist at Washington University in St. Louis, with cloning, "there'd be no need for men."

But on a more serious note, Dr. Stanley Hauerwas, a divinity professor at Duke University, said that those who wanted to clone "are going to sell it with wonderful benefits" for medicine and animal husbandry. But he said he saw "a kind of drive behind this for us to be our own creators."

Dr. Kevin FitzGerald, a Jesuit priest and a geneticist at Loyola University in Maywood, Ill., cautioned that people might not understand clones. While a clone would be an identical, but much younger, twin of the adult, people are more than just the sum of their genes. A clone of a human being, he said, would have a different environment than the person whose DNA it carried and so would have to be a different person. It would even have to have a different soul, he added.

The cloning was done by Dr. Ian Wilmut, a 52-year-old embryologist at the Roslin Institute in Edinburgh. Dr. Wilmut announced on Saturday that he had replaced the genetic material of a sheep's egg with the DNA from an adult sheep and created a lamb that is a clone of the adult. He is publishing his results in the British journal *Nature* on Thursday.

While other researchers had previously produced genetically identical animals by dividing embryos soon after they had been formed by eggs and sperm, Dr. Wilmut is believed to be the first to create a clone using DNA from an adult animal. Until now, scientists believed that once adult cells had differentiated—to become skin or eye cells, for example—their DNA would no longer be usable to form a complete organism.

Dr. Wilmut reported that as a source of genetic material, he had used udder, or mammary, cells from a 6-year-old adult sheep. The cells were put into tissue culture and manipulated to make their DNA become quiescent. Then Dr. Wilmut removed the nucleus, containing the genes, from an egg cell taken from another ewe. He fused that egg cell with one of the adult udder cells.

> *. . . Dr. Wilmut said he wanted to create animals that could be used for medical research, and he dismissed the notion of cloning humans.*

When the two cells merged, the genetic material from the adult took up residence in the egg and directed it to grow and divide. Dr. Wilmut implanted the developing embryo in a third sheep, who gave birth to a lamb that is a clone of the adult that provided its DNA. The lamb, named Dolly, was born in July and seems normal and healthy, Dr. Wilmut said.

In an interview, Dr. Wilmut said he wanted to create animals that could be used for medical research, and he dismissed the notion of cloning humans. "There is no reason in principle why you couldn't do it," he said. But he added, "All of us would find that offensive."

Yet others said that might be too glib. "It is so typical for scientists to say they are not thinking about the implications of their work," said Dr. Lee Silver, a biology professor at Princeton University. Perhaps, he added, "the only way they can validate what they are doing is to say they are just doing it in sheep."

Few experts think that sheep or other farm animals would be the only animals to be cloned. While cloning people is illegal in Britain and several other countries, John Robertson, a law professor at the University of Texas at Austin who studies reproductive rights and bioethics, said there were no laws against it in the United States.

If such a law was passed, Dr. Silver said, doctors could set up clinics elsewhere to offer cloning. "There's no way to stop it," Dr. Silver said. "Borders don't matter."

Dr. Ronald Munson, an ethicist at the University of Missouri at St. Louis, said the cloning itself was relatively simple. "This technology is not, in principle, policeable," he said. "It doesn't require the sort of vast machines that you need for atom-smashing. These are relatively standard labs. That's the amazing thing about all this biotechnology. It's fundamentally quite simple."

One immediate implication of cloning, Dr. Silver said, would be for genetic engineering: custom-tailoring genes. Currently, scientists are unable to take a gene and simply add it to cells. The process of adding genes is so inefficient that researchers typically have to add genes to a million cells to find one that takes them up and uses them properly. That makes it very difficult to add genes to an embryo—or a person—to correct a genetic disease or genetically enhance a person, Dr. Silver said. But now, "it all becomes feasible," he said.

After adding genes to cells in the laboratory, scientists could fish out the one cell in a million with the right changes and use it to clone an animal—or a person. "All of a sudden, genetic engineering is much, much easier," Dr. Silver said.

> *One immediate implication of cloning . . . would be for genetic engineering: custom-tailoring genes.*

Dr. Wilmut is hoping that the genes for pharmacologically useful proteins could be added to sheep mammary cells and that the best cells could be used for cloning. The adult cloned sheep would produce the proteins in their milk, where they could be easily harvested.

Because cloning had been considered so far-fetched, scientists had discouraged ethicists from dwelling on its implications, said Dr. Daniel Callahan, a founder of the Hastings Center, one of the first ethics centers.

In the early 1970's, "there was an enormous amount of discussion about cloning," Dr. Callahan said, and ethicists mulled over the frightening implications. But scientists dismissed these discussions as idle speculation about impossible things, Dr. Callahan recalled, and urged ethicists not to dwell on the topic.

"A lot of scientists got upset," Dr. Callahan said. "They said that this is exactly the sort of thing that brings science into bad repute and you people should stop talking about it."

In the meantime, however, cloning had captured the popular imagination. In his 1970 book, *Future Shock,* Alvin Toffler speculated that "cloning would make it possible for people to see themselves anew, to fill the world with twins of themselves."

Woody Allen's 1973 movie *Sleeper* involved a futuristic world whose leader had left behind his nose for cloning purposes. Mr. Allen played a character charged with cloning to bring the leader back. A later movie, *The Boys from Brazil,* released in 1978, involved a Nazi scheme to clone multiple Hitlers. That same year, a science writer, David Rorvik, published a book, *In His Image: The Cloning of a Man,* that purported to be the true story of a wealthy man who had secretly had himself cloned but was found to be a hoax.

But gradually, the notion disappeared from sight, kept alive only in the animal husbandry industry, where companies saw a huge market for cloned animals and where the troubling ethical implications of cloning could be swept aside.

Now these questions are back to haunt ethicists and theologians.

Clones of animals, Dr. FitzGerald said, might sound appealing—scientists could clone the buttery Kobe beef cattle or the meatiest pigs, for example. But these cloned creatures would also share an identical susceptibly to disease, he cautioned. An entire cloned herd could be wiped out overnight if the right virus swept through it.

Dr. Fitzgerald wondered if people would actually try to clone themselves. "Because we have all this technology and we have this ability," he said, "we can spin off these fantasies. But that doesn't mean we'd do it. It would be going against everything we desire for the human race."

Others are less sure. Mr, Robertson can envision times when cloning might be understandable. Take the case of a couple whose baby was dying and who wanted, literally, to replace the child. Mr. Robertson does not think that would be so reprehensible.

Cloning might also be attractive to infertile couples who want children and who "want to be sure that whatever offspring they have has good genes," Mr. Robertson said.

Of course, there are legal issues, Mr. Robertson said, like the issue of consent. "Would the person being cloned have an intellectual property right or basic human right to control their DNA?" he asked. If the person did, and consented to the cloning, would cloning be procreation, as it is now understood?

Mr. Robertson thinks not. After all, he said, "replication is not procreation."

For Cloning Pioneer, Biotechnology Holds Promise of Medical Gains[2]

BY MARJORIE MILLER
LOS ANGELES TIMES, APRIL 9, 2000

Dolly, the world's most famous sheep, is indisposed, having just given birth for the third time in as many years.

Ian Wilmut, laboratory father of the first mammal cloned from an adult cell, is somewhat more available, seated next to a computer that is flashing the time. The digital seconds fly off the screen at a heart-palpitating rate.

Time is of the essence for both members of the cutting-edge Roslin Institute since Dolly heralded a biological revolution three years ago. In the sheep's case, the concern is with signs of premature aging that scientists are unable to explain. She looks healthy to an examining vet, but under the microscope, a part of her chromosomes is seen to have shortened more than is normal for her years.

What does it mean and will it lead to an early death? No one knows yet.

Wilmut's preoccupation, meanwhile, is with all of the pressure that Dolly's fame has put on his time: the speeches, interviews and inquiries competing with demands for more research to push the frontiers of biotechnology even further.

More than most scientists, the bald and bespectacled embryologist has had to respond to the enormous moral questions and brave-new-world fears raised by his groundbreaking work in genetic engineering and cloning: Is it right to tamper with nature? Are we headed for human cloning? Should companies be allowed to patent genes and DNA, the basis of life?

At the same time, he has fielded heartbreaking telephone calls from parents wondering if it is possible to clone a young child who died. As the son of a diabetic who was blinded and crippled by the disease, Wilmut feels an urgency to discover biological cures to such illnesses. He knows too that thousands of kidney and heart patients are desperate for scientists to develop an animal whose organs are fit for transplanting to humans.

"From the first time man put a sharp stone to a stick, he could use this as a tool to eat with or as a weapon to kill people," Wilmut

---

2. Article by Marjorie Miller from the *Los Angeles Times* April 9, 2000. Copyright © *The Los Angeles Times*. Reprinted with permission.

said. "It's not a dilemma—it's a fact of life. We have to get used to making these choices."

Wilmut is taking time to talk because he feels strongly that the public must be educated about the new science and technology to make informed policy decisions. To that end, he wrote *The Second Creation* with Keith Campbell, his scientific partner in cloning Dolly, and author Colin Tudge.

The book, to be published in the United States in June, is textbook dense and complicated. Wilmut, on the other hand, is pleasantly plain-spoken.

> *One of the most persistent misconceptions of this technology is that clones are identical, Wilmut says.*

"We want to promote discussion of the potential uses of the technology. We profoundly believe that these are social decisions that should not be left to the scientists, the companies, the doctors and the patients, who are all too involved," Wilmut said.

Before Dolly was cloned by destroying the nucleus of an adult sheep cell and replacing it with another nucleus, scientists believed that such a procedure was "biologically impossible" in mammals. Dolly rendered that term obsolete and showed that future limits on such technology would be determined not by biology but by ethics.

The moral and emotional complexities of biotechnology were driven home to Wilmut immediately after the announcement of Dolly's birth: The German magazine *Der Spiegel* accompanied its report with a cover illustration of a regiment of cloned Hitlers.

One of the most persistent misconceptions of this technology is that clones are identical, Wilmut says. They are not, as was proved by Roslin's four cloned rams—Cedric, Cecil, Cyril and Tuppence, who are genetically identical but different in size, color and temperament.

Here again, scientists do not know why. One possibility is that the nuclei from which they were created were placed in egg cells from four ewes and thus developed in distinct environments.

"I often tell people that a genetic copy of Saddam Hussein might well turn out to be a nice guy," he said.

No regiment of Hitlers; no army of Einsteins either.

Wilmut finds the idea of human cloning abhorrent, although he is only too aware of his role in making it possible. There is no medical justification for attempting human cloning, he says. He believes that it would be "grossly irresponsible" to consider trying to clone a human, given scientists' lack of expertise in the field and the lack of knowledge about what it means to be a clone.

Most cloned sheep embryos die, many of them after having developed serious deformities. Only 1% to 4% of reconstituted embryos survive to become live offspring, Wilmut says. Of those, 20% die shortly after birth—four times as many as in a naturally bred flock.

Then, there are the unknowns about life as a clone. Dolly is an apparently healthy, if spoiled, ewe who has delivered normal offspring and who would ordinarily live about 12 years. What do the shortened ends of her chromosomes mean? One possibility is that she will be more vulnerable to tumors than are normal sheep.

Wilmut has three grown children of his own and a 1-year-old grandson whom he clearly adores. He can understand a parent's willingness to try anything to save—or replace—a child. But he wouldn't push the limits of biotechnology even for his own family.

"The image is that you bring a child back. You don't do that. My daughter is a flute player and active Christian. A copy of her might not be either one of those things," he said, adding that to try to make a baby into the person someone else was would be a tragedy.

Wilmut finds the idea of human cloning abhorrent, although he is only too aware of his role in making it possible.

So what does Wilmut consider to be the legitimate pursuits of biotechnology? Improving livestock, certainly. Cell therapy, substituting healthy cells for diseased ones to treat the likes of diabetes or Parkinson's. Maybe even organ transplants, although Wilmut says they are a long way off despite the recent birth of cloned quintuplet pigs—the most likely species to become organ donors.

Scientists already have created animals fitted with human genes to produce proteins in their milk that can be used for treating diseases such as emphysema, cystic fibrosis and hemophilia. The therapies are still three to five years from coming on the market, Wilmut says.

PPL Therapeutics, the commercial pharmaceutical company located on the grounds of the government-funded Roslin Institute, announced last month that it had created the first cloned pigs, offering hope of farming animals for organs such as hearts and kidneys to transplant into humans.

The pigs must be genetically altered so that their organs will not be rejected by a human body, and countless tests must be done to ensure that dormant viruses would not be passed on to humans. The company says human trials could begin in as little as four years. Wilmut says six to 10 years is more realistic.

And then what? Human cloning is illegal in the U.S. and Britain, but is it the logical next step? Will there be a commercial drive for human cloning one day?

Most of the investment in biotechnology research comes from companies hoping to profit from cellular therapies and organs. The Roslin Institute and the government agencies that funded its cloning research own the patents for the technology that led to Dolly.

Roslin formed a commercial affiliate, Bio-Med, that merged with the California-based Geron Corp., which is licensed to exploit the cloning research for its program to create replacement tissues and organs. Wilmut volunteers that he has stock in Geron and stands to earn a lot of money from new breakthroughs in biotechnology.

Thirty years ago, scientists like Wilmut would never have mentioned the word "patent," he said. "Now, I have a contractual obligation to think of patents."

Many people believe that these technologies should not be patented because they revolve around the makeup of human life. Others worry that the patents for controversial procedures are being given to private companies governed by commercial, rather than ethical interests.

Will there be a commercial drive for human cloning one day?

President Clinton and British Prime Minister Tony Blair have called on companies to give free access to discoveries on the makeup of the human genetic code. People have 60,000 to 80,000 genes, but scientists know what only a few thousand of them do. The faster the genes are identified, the more likely new drugs can be developed to treat diseases.

Wilmut says governments need to clarify what can and cannot be patented. He believes that it is inappropriate to patent a list of genes but reasonable to patent the inventive use of a gene for producing proteins needed to treat diseases. A patent ensures both a return for investments in biotechnology and quick dissemination of scientific information, he argues.

Scientists can then build on the information to "take humanity into the age of biological control," as Wilmut and Campbell state in their book.

But that does not answer the question of whether all of this is likely to lead to human cloning.

Wilmut, a middle-class Englishman who looks like a high school teacher, is an ordinary man who has done something extraordinary. He responds carefully that the technology for human cloning does not exist yet and that he hopes scientists will not go down this road. But that is probably a futile hope, he acknowledges.

"Is it inevitable that cloning will be used to copy a person? No, it's not inevitable," he said. "It all depends on what societies decide they want to do. Is it likely it will happen? Yes."

Of Headless Mice ... And Men[3]

The Ultimate Cloning Horror: Human Organ Farms

BY CHARLES KRAUTHAMMER
TIME, JANUARY 19, 1998

Last year Dolly the cloned sheep was received with wonder, titters and some vague apprehension. Last week the announcement by a Chicago physicist that he is assembling a team to produce the first human clone occasioned yet another wave of Brave New World anxiety. But the scariest news of all—and largely overlooked—comes from two obscure labs, at the University of Texas and at the University of Bath. During the past four years, one group created headless mice; the other, headless tadpoles.

For sheer Frankenstein wattage, the purposeful creation of these animal monsters has no equal. Take the mice. Researchers found the gene that tells the embryo to produce the head. They deleted it. They did this in a thousand mice embryos, four of which were born. I use the term loosely. Having no way to breathe, the mice died instantly.

Why then create them? The Texas researchers want to learn how genes determine embryo development. But you don't have to be a genius to see the true utility of manufacturing headless creatures: for their organs—fully formed, perfectly useful, ripe for plundering.

Why should you be panicked? Because humans are next. "It would almost certainly be possible to produce human bodies without a forebrain," Princeton biologist Lee Silver told the *London Sunday Times*. "These human bodies without any semblance of consciousness would not be considered persons, and thus it would be perfectly legal to keep them 'alive' as a future source of organs."

"Alive." Never have a pair of quotation marks loomed so ominously. Take the mouse-frog technology, apply it to humans, combine it with cloning, and you become a god: with a single cell taken from, say, your finger, you produce a headless replica of yourself, a mutant twin, arguably lifeless, that becomes your own personal, precisely tissue-matched organ farm.

---

3. Article by Charles Krauthammer from *Time Magazine* January 19, 1998. Copyright © Time-Life Syndication. Reprinted with permission.

There are, of course, technical hurdles along the way. Suppressing the equivalent "head" gene in man. Incubating tiny infant organs to grow into larger ones that adults could use. And creating artificial wombs (as per Aldous Huxley), given that it might be difficult to recruit sane women to carry headless fetuses to their birth/death.

It won't be long, however, before these technical barriers are breached. The ethical barriers are already cracking. Lewis Wolpert, professor of biology at University College, London, finds producing headless humans "personally distasteful" but, given the shortage of organs, does not think distaste is sufficient reason not to go ahead with something that would save lives. And Professor Silver not only

---

There is no grosser corruption of biotechnology than creating a human mutant and disemboweling it at our pleasure for spare parts.

---

sees "nothing wrong, philosophically or rationally," with producing headless humans for organ harvesting; he wants to convince a skeptical public that it is perfectly O.K.

When prominent scientists are prepared to acquiesce in—or indeed encourage—the deliberate creation of deformed and dying quasi-human life, you know we are facing a bioethical abyss. Human beings are ends, not means. There is no grosser corruption of biotechnology than creating a human mutant and disemboweling it at our pleasure for spare parts.

The prospect of headless human clones should put the whole debate about "normal" cloning in a new light. Normal cloning is less a treatment for infertility than a treatment for vanity. It is a way to produce an exact genetic replica of yourself that will walk the earth years after you're gone.

But there is a problem with a clone. It is not really you. It is but a twin, a perfect John Doe Jr., but still a junior. With its own independent consciousness, it is, alas, just a facsimile of you.

The headless clone solves the facsimile problem. It is a gateway to the ultimate vanity: immortality. If you create a real clone, you cannot transfer your consciousness into it to truly live on. But if you create a headless clone of just your body, you have created a ready source of replacement parts to keep you—your consciousness—going indefinitely.

Which is why one form of cloning will inevitably lead to the other. Cloning is the technology of narcissism, and nothing satisfies narcissism like immortality. Headlessness will be cloning's crowning achievement.

The time to put a stop to this is now. Dolly moved President Clinton to create a commission that recommended a temporary ban on human cloning. But with physicist Richard Seed threatening to clone humans, and with headless animals already here, we are past the time for toothless commissions and meaningless bans.

Clinton banned federal funding of human-cloning research, of which there is none anyway. He then proposed a five-year ban on cloning. This is not enough. Congress should ban human cloning now. Totally. And regarding one particular form, it should be draconian: the deliberate creation of headless humans must be made a crime, indeed a capital crime. If we flinch in the face of this high-tech barbarity, we'll deserve to live in the hell it heralds.

The Judeo-Christian Case against Human Cloning[4]

BY STEPHEN G. POST
AMERICA, JUNE 21-28, 1997

The very idea of cloning tends to focus on the physiological substrate, not on the journey of life and our responses to it.

For purposes of discussion, I will assume that the cloning of humans is technologically possible. This supposition raises Einstein's concern: "Perfection of means and confusion of ends seems to characterize our age." Public reaction to human cloning has been strongly negative, although without much clear articulation as to why. My task is the Socratic one of helping to make explicit what is implicit in this uneasiness.

Some extremely hypothetical scenarios might be raised as if to justify human cloning. One might speculate, for example: If environmental toxins or pathogens should result in massive human infertility, human cloning might be imperative for species survival. But in fact recent claims about increasing male infertility worldwide have been found to be false. Some apologists for human cloning will insist on other strained "What if's." "What if" parents want to replace a dead child with an image of that child? "What if" we can enhance the human condition by cloning the "best" among us?

I shall offer seven unhypothetical criticisms of human cloning, but in no particular priority. The final criticism, however, is the chief one to which all else serves as preamble.

1. The Newness of Life. Although human cloning, if possible, is surely a novelty, it does not corner the market on newness. For millennia mothers and fathers have marveled at the newness of form in their newborns. I have watched newness unfold in our own two children, wonderful blends of the Amerasian variety. True, there probably is, as Freud argued, a certain narcissism in parental love, for we do see our own form partly reflected in the child, but, importantly, never entirely so. Sameness is dull, and as the French say, *Vive la difference.* It is possible that underlying the mystery of this newness of form is a creative wisdom that we humans will never quite equal.

This concern with the newness of each human form (identical twins are new genetic combinations as well) is not itself new. The scholar of constitutional law Laurence Tribe pointed out in 1978, for example, that human cloning could "alter the very meaning of humanity." Specifically, the cloned person would be "denied a sense of uniqueness." Let us remember that there is no strong analogy between human cloning and natural identical twinning, for in the latter case there is still the blessing of newness in the newborns, though they be two or more. While identical twins do occur naturally and are unique persons, this does not justify the temptation to impose external sameness more widely.

Sidney Callahan, a thoughtful psychologist, argues that the random fusion of a couple's genetic heritage "gives enough distance to allow the child also to be seen as a separate other," and she adds that the egoistic intent to deny uniqueness is wrong because ultimately depriving. By having a different form from that of either parent, I am visually a separate creature, and this contributes to the moral purpose of not reducing me to a mere copy utterly controlled by the purposes of a mother or father.

Perhaps human clones will not look exactly alike anyway, given the complex factors influencing genetic imprinting, as well as environmental factors affecting gene expression. But they will look more or less the same, rather than more or less different.

> *Surely no scientist would doubt that genetic diversity produced by procreation between a man and a woman will always be preferable to cloning . . . .*

Surely no scientist would doubt that genetic diversity produced by procreation between a man and a woman will always be preferable to cloning, because procreation reduces the possibility for species annihilation through particular diseases or pathogens. Even in the absence of such pathogens, cloning means the loss of what geneticists describe as the additional hybrid vigor of new genetic combinations.

2. Making Males Reproductively Obsolete. Cloning requires human eggs, nuclei and uteri, all of which can be supplied by women. This makes males reproductively obsolete. This does not quite measure up to Shulamith Firestone's notion that women will only be able to free themselves from patriarchy through the eventual development of the artificial womb, but of course, with no men available, patriarchy ends—period.

Cloning, in the words of Richard McCormick, S.J., "would involve removing insemination and fertilization from the marriage relationship, and it would also remove one of the partners from the entire process." Well, removal of social fatherhood is already a fait accompli in a culture of illegitimacy chic, and one to which some fertility clinics already marvelously contribute through artificial insemination by donor for single women. Removing male impregnators from the procreative dyed would simply drive the nail into the coffin of fatherhood, unless one thinks that biological and social fatherhood are utterly disconnected. Social fatherhood would still be possible in a world of clones, but this will lack the feature of participation in a continued biological lineage that seems to strengthen social fatherhood in general.

3. Under My Thumb: Cookie Cutters and Power. It is impossible to separate human cloning from concerns about power. There is the power of one generation over the external form of another, imposing the vicissitudes of one generation's fleeting image of the good upon the nature and destiny of the next. One need only peruse the innumerable texts on eugenics written by American geneticists in the 1920's to understand the arrogance of such visions.

> *It is impossible to separate human cloning from concerns about power.*

One generation always influences the next in various ways, of course. But when one generation can, by the power of genetics, in the words of C. S. Lewis, "make its descendants what it pleases, all men who live after it are the patients of that power." What might our medicalized culture's images of human perfection become? In Lewis' words again, "For the power of Man to make himself what he pleases means, as we have seen, the power of some men to make other men what they please."

A certain amount of negative eugenics by prenatal testing and selective abortion is already established in American obstetrics. Cloning extends this power from the negative to the positive, and it is therefore even more foreboding.

This concern with overcontrol and overpower may be overstated because the relationship between genotype and realized social role remains highly obscure. Social role seems to be arrived at as much through luck and perseverance as anything else, although some innate capacities exist as genetically informed baselines.

4. Born to Be Harvested. One hears regularly that human clones would make good organ donors. But we ought not to presume that anyone wishes to give away body parts. The assumption that the clone would choose to give body parts is completely unfounded. Forcing such a harvest would reduce the clone to a mere object for the use of others. A human person is an individual substance of a rational nature not to be treated as object, even if for one's own nar-

cissistic gratification, let alone to procure organs. I have never been convinced that there are any ethical duties to donate organs.

5. The Problem of Mishaps. Dolly the celebrated ewe represents one success out of 277 embryos, nine of which were implanted. Only Dolly survived. While I do not wish to address here the issue of the moral status of the entity within the womb, suffice it to note that in this country there are many who would consider proposed research to clone humans as far too risky with regard to induced genetic defects. Embryo research in general is a matter of serious moral debate in the United States, and cloning will simply bring this to a head.

As one recent British expert on fertility studies writes, "Many of the animal clones that have been produced show serious developmental abnormalities, and, apart from ethical considerations, doctors would not run the medico-legal risks involved."

6. Sources of the Self. Presumably no one needs to be reminded that the self is formed by experience, environment and nurture. From a moral perspective, images of human goodness are largely virtue-based and therefore characterological. Aristotle and Thomas Aquinas believed that a good life is one in which, at one's last breath, one has a sense of integrity and meaning. Classically the shaping of human fulfillment has generally been a matter of negotiating with frailty and suffering through perseverance in order to build character. It is not the earthen vessels, but the treasure within them that counts. A self is not so much a genotype as a life journey. Martin Luther King Jr. was getting at this when he said that the content of character is more important than the color of skin.

The very idea of cloning tends to focus images of the good self on the physiological substrate, not on the journey of life and our responses to it, some of them compensations to purported "imperfections" in the vessel. The idea of the designer baby will emerge, as though external form is as important as the inner self.

7. Respect for Nature and Nature's God. In the words of Jewish bioethicist Fred Rosner, cloning goes so far in violating the structure of nature that it can be considered as "encroaching on the Creator's domain." Is the union of sex, marriage, love and procreation something to dismiss lightly?

Marriage is the union of female and male that alone allows for procreation in which children can benefit developmentally from both a mother and father. In the Gospel of Mark, Jesus draws on ancient Jewish teachings when he asserts, "Therefore what God has joined together, let no man separate." Regardless of the degree of extendedness in any family, there remains the core nucleus: wife, husband and children. Yet the nucleus can be split by various cultural forces (e.g., infidelity as interesting, illegitimacy as chic), poverty, patriarchal violence and now cloning.

A cursory study of the Hebrew Bible shows the exuberant and immensely powerful statements of Genesis 1, in which a purposeful, ordering God pronounces that all stages of creation are "good."

> *Embryo research in general is a matter of serious moral debate in the United States, and cloning will simply bring this to a head.*

The text proclaims, "So God created humankind in his image, in the image of God he created them, male and female he created them" (Gen. 1:27). This God commands the couple, each equally in God's likeness, to "be fruitful and multiply." The divine prototype was thus established at the very outset of the Hebrew Bible: "Therefore a man leaves his father and his mother and clings to his wife, and they become one flesh" (Gen. 2:24).

The dominant theme of Genesis I is creative intention. God creates, and what is created procreates, thereby ensuring the continued presence of God's creation. The creation of man and woman is good in part because it will endure. Catholic natural law ethicists and Protestant proponents of "orders of creation" alike find divine will and principle in the passages of Genesis 1.

A major study on the family by the Christian ethicist Max Stackhouse suggests that just as the pre-Socratic philosophers discovered still valid truths about geometry, so the biblical authors of Chapters One and Two of Genesis "saw something of the basic design, purpose, and context of life that transcends every sociohistorical epoch." Specifically, this design includes "fidelity in communion" between male and female oriented toward "generativity" and an enduring family the precise social details of which are worked out in the context of political economies.

Christianity appropriated the Hebrew Bible and had its origin in a Jew from Nazareth and his Jewish followers. The basic contours of Christian thought on marriage and family therefore owe a great deal to Judaism. These Hebraic roots that shape the words of Jesus stand within Malachi's prophetic tradition of emphasis on inviolable monogamy. In Mk. 10:2-12 we read:

> The Pharisees approached and asked, "Is it lawful for a husband to divorce his wife?" They were testing him. He said to them in reply, "What did Moses command you?" They replied, "Moses permitted him to write a bill of divorce and dismiss her." But Jesus told them, "Because of the hardness of your hearts he wrote you this commandment. But from the beginning of creation, 'God made them male and

female. For this reason a man shall leave his father and mother (and be joined to his wife), and the two shall become one flesh.' So they are no longer two but one flesh. Therefore what God has joined together, no human being must separate." In the house the disciples again questioned him about this. He said to them, "Whoever divorces his wife and marries another commits adultery against her; and if she divorces her husband and marries another, she commits adultery."

Here Jesus quotes Gen. 1:27 ("God made them male and female") and Gen. 2:24 ("the two shall become one flesh").

Christians side with the deep wisdom of the teachings of Jesus, manifest in a thoughtful respect for the laws of nature that reflect the word of God. Christians simply cannot and must not underestimate the threat of human cloning to unravel what is both naturally and eternally good.

V.

Rebuilding the
Human Body

Editor's Introduction

The following lines are familiar to any fan of 1970s television: "We can rebuild him. We have the technology. We have the capability to make the world's first Bionic man. Steve Austin will be that man. Better than he was before. Better . . . stronger . . . faster." Many of those who watched *The Six Million Dollar Man* as children and adolescents dreamed of creating their own bionic man or woman. They imagined taking a broken body and refashioning it into the human equivalent of a souped-up sports car—someone who was better, stronger, and faster than the average person. Some of those young people grew up to become today's medical researchers and biotechnology experts who strive to improve the human condition through the manufacture of replacement parts for those in need of transplants or prostheses. As Section V illustrates, their goal is to help rebuild human bodies, to make them as good as they were before or, possibly, even better. Now, thanks to their efforts, in some cases we *do* have the technology to do just that. With the shortage of transplantable organs nearing the crisis stage, many of the innovations discussed in the following articles could be life savers, while others could offer hope for the disabled and a means to improve the quality of their lives.

The article which begins the section, Vincent Kiernan's "Can Bioengineering Create a Human Heart?" written for the *Chronicle of Higher Education*, discusses the worldwide collaboration called the Living Implants From Engineering Initiative, or LIFE. The ambitious goal of the researchers in LIFE is to manufacture, out of human cells, a quantity of such vital organs as hearts, livers, and kidneys. A human heart produced in this way could then receive its nutrients from the individual's own blood. As Kiernan explains, having a supply of such artificial hearts on hand would not only save the lives of those currently on the waiting list for transplants, but, according to the head of the LIFE project, it would also save the money that would otherwise be needed to treat those hospitalized with serious cardiovascular disease.

Another kind of transplant has raised eyebrows since it was first performed in the 1960s, and that is the subject of Malcom W. Browne's *New York Times* article called "From Science Fiction to Science: 'The Whole Body Transplant.'" Browne explains how Dr. Robert J. White at the Case Western Reserve University School of Medicine has been encouraged by the successful transplantations of the heads of monkeys, cats, and dogs and now wants to attempt the same procedure on human beings, claiming that quadruplegics would particularly benefit from the "whole body transplant." In his article, Browne explores the feasibility of such a transplant and discusses the "unsettling" physiological and philosophical implications of this particular kind of procedure.

The shortage of organ donors in the U.S. and throughout the world has prompted some physicians to attempt animal transplants in their most desperately ill patients, and this is the subject of the next article, Amy Otchet's "Animal Transplants: Safe or Sorry?" written for the *UNESCO Courier*. Otchet addresses a number of the benefits and pitfalls of this particular type of surgery, called xenotransplantation, including the possibility of the patient contracting animal-specific diseases and fears of epidemics on the order of HIV or so-called "mad cow disease." As Otchet explains, while many scientists continue to experiment with ways to genetically engineer animals, such as pigs, with an immune system similar to that of human beings, thereby making xenotransplantation safer for patients, other researchers remain cautious or "unconvinced" about the feasibility of such transplants.

In the article which follows, entitled "Mind Over Matter," written by Demaris Christensen for *Science News*, reports on one of the most innovative means of aiding quadruplegics: prostheses controlled by the brain's electrical impulses. Despite the enormous difficulties scientists have found in harnessing the neurological signals sent by the human brain, Christensen says, researchers remain hopeful that within a decade or two, implanted electrodes will enable those paralyzed by injury or diseases, such as ALS (commonly known as Lou Gehrig's disease), to communicate via computer and operate a wheelchair or prosthetic arm. Until such implants are perfected, external devices may be able to achieve the same objective, opening a range of possibilities for those currently suffering from nearly total or complete paralysis.

The section concludes with "Cyborg 1.0," Kevin Warwick's first-person account of his own experience with a silicon chip implant written for *Wired*. In this piece, which details his unusual experiment in cybernetics, Warwick describes feeling a close emotional connection with the computer with which he communicates via the chip. He also speculates on the future applications of this new technology, including the development of cyberdrugs and brain implants which will allow all human beings to engage in "thought communication," thereby obviating the need for spoken language. He then explains how his wife, Irena, could soon become the second cyborg and join him in exploring life as a "superhuman."

Can Bioengineering Create a Human Heart?[1]

By Vincent Kiernan
Chronicle of Higher Education, February 19, 1999

Michael V. Sefton has another solution for the thousands of people around the world who desperately need human transplants to replace their diseased hearts: A huge supply of new, implantable human hearts grown from scratch with biotechnology.

"The idea is to have hearts on a shelf," ready for implantation into a patient whenever the need arises, said Mr. Sefton a professor of chemical engineering and applied chemistry at the University of Toronto and a researcher at its Institute of Biomedical Engineering.

Mr. Sefton is trying to organize bioengineering colleagues around the world into a massive collaboration called the Living Implants From Engineering Initiative, or LIFE. The project—which is only in the process of being organized and has not yet made any scientific progress—aims at using biotechnology to produce a supply of implantable human vital organs, such as hearts, livers, and kidneys. Unlike mechanical hearts under development by other researchers, a bioengineered heart would be made of human cardiac cells that have been grown and modified with biotechnology processes.

Developing a bioengineered heart would take at least a decade and cost between $1 billion and $5 billion, Mr. Sefton told last month's annual meeting of the American Association for the Advancement of Science here. Even some of the researchers with whom he is working think he's optimistic with that schedule.

A Godsend for Many

But a bioengineered heart would be a godsend for many. In 1997, the most recent year for which figures are available, 4,331 people in the United States died while on the waiting list for an organ transplant, including 773 who died while waiting for hearts, according to the United Network for Organ Sharing, a non-profit

---

1. Article by Vincent Kiernan from *The Chronicle of Higher Education* February 19, 1999. Copyright © *The Chronicle of Higher Education*. Reprinted with permission. This article may not be posted, published, or further distributed without permission from *The Chronicle*.

corporation in Richmond, Va., which coordinates organ transplants in the United States.

The LIFE project even could save society money, Mr. Sefton said. Although developing a bioengineered heart would cost a substantial sum, transplanting bioengineered hearts into those who are ill would avoid the need for spending billions more for medical care for those dying of cardiovascular disease, he told the science conference. He estimates that implanting a bioengineered heart would cost about the same as transplanting a heart from a human donor.

Mr. Sefton's proposal has attracted substantial interest among bioengineering researchers. Scientists from 13 institutions attended an initial planning meeting for the LIFE Initiative last

---

... transplanting bioengineered hearts into those who are ill would avoid the need for spending billions more for medical care for those dying of cardiovascular disease ....

June in Toronto. And since then, the roster of interested researchers has grown to represent dozens of universities.

"They are excited about this area and any involvement that would help people," said Robert S. Langer, a professor of chemical and biomedical engineering at the Massachusetts Institute of Technology and one of the collaborators in the LIFE Initiative.

But the project itself is still in a nascent stage and has not yet produced any scientific results—and not even a clear blueprint for producing organs.

Mr. Sefton said that the initiative might not tackle hearts first. Rather, the scientists might initially try to develop a bioengineered liver or kidney. Those organs, which remove waste and toxins from the blood, are simpler than a heart to duplicate, he said.

Public-relations Value

But when it comes to biomedical research, the heart has unparalleled public-relations value in attracting public attention and funds, which is why Mr. Sefton emphasizes the heart in his public comments.

"To be honest, that's part of it," he acknowledged in an interview.

There are several possible approaches for developing the bioengineered organs, Mr. Sefton said. One option would be to create a "scaffold" of a biodegradable polymer in the shape of the desired organ. Researchers would then attach cells to that scaffold. As the cells matured, they would take on the scaffold's shape, and the scaffold itself would gradually decompose.

Another possibility, according to Mr. Sefton, is to use "stem cells," or cells that have not yet been programmed to become specific types of cells, such as heart or muscle or bone cells. In an embryo, stem cells develop into the different types of cells found in various organs in the body. The researchers would use genetic-engineering techniques to induce the stem cells to evolve into whatever organ they were seeking to produce for implantation into humans.

The use of stem cells in research has become contentious as biomedical researchers have begun to develop techniques to keep them alive in the laboratory. Last month, the N.I.H. ruled that a ban on using federal funds for research with human embryos did not cover research with stem cells.

A third approach to producing bioengineered organs, Mr. Sefton said, is to build parts of an organ, using scaffolding or stem cells, depending on which technique was better suited for a particular component. The parts would be assembled into a finished organ in the operating room, just before implantation. For example, a heart's valves might be manufactured separately from the rest of the heart and attached to the heart in the operating room.

But each of those approaches may be difficult to achieve. Mr. Sefton said, for example, that researchers did not yet fully understand how stem cells are triggered into becoming cardiac cells. They do not even know how to keep thick pieces of cardiac tissue alive in the laboratory, he added.

Avoiding Rejection

Researchers also would have to find a way to prevent the body's immune system from rejecting the bioengineered heart.

One possibility, said Mr. Sefton, is to modify the cells in the bioengineered organ so that they do not contain protein markers that would trigger a rejection. Another approach, he said, is to find ways to suppress the body's immune response to the implanted organ.

The researchers still are trying to determine how to organize themselves. One possibility, he and others say, is to pattern their work after the Human Genome Project, a huge effort financed by the National Institutes of Health and the U.S. Energy Department to determine the genetic sequence of human DNA. The two agencies coordinate the project and provide funds to hundreds of researchers in universities, industry, and government laboratories.

In that vein, said Mr. Sefton, the LIFE Initiative might form an organization that would serve as the central coordinating body for the bioengineering research at many institutions, distributing funds to each participant. Or N.I.H. itself might play that coordinating role.

Foundations and Individual Donors

John Watson, acting deputy director of N.I.H.'s National Heart, Lung, and Blood Institute, said that the N.I.H financed about 30 small research projects in tissue engineering and that great strides had been made in the technology. But he said it would be premature to consider any proposal for the government to bankroll a large-scale program in bioengineered organs. "I would think that investors in something like this would like to see some evidence of its feasibility," he said.

> *"I would think that investors in something like this would like to see some evidence of its feasibility."*—John Watson, acting deputy director of N.I.H.'s National Heart, Lung, and Blood Institute

Later this year, Mr. Sefton and some of the researchers in the initiative will begin to approach foundations and individual donors for seed money to start the project and to conduct initial experiments.

At the science association's meeting, he joked that the heart's container could bear a Nike "swoosh," as a way to attract funds from the sneaker manufacturer.

But in the early stages, he said, the researchers will eschew money from industry, for fear that a company that provided seed money to the project might some day claim commercial rights to all the technology and processes that the LIFE initiative might produce.

Even without corporate involvement, Mr. Sefton said, the LIFE initiative will face the difficult task of determining how universities will share in any commercial revenue that the project eventually generates.

"Each university is going to think it owns the rights to that product," said Mr. Sefton. Consequently, he and other organizers are trying to figure out a way to define, in advance, how ownership of the LIFE Initiative's products will be divided among participating universities.

As the program matures, however, it will need corporate money—and access to corporate biomedical technologies to perfect and produce whatever products the initiative yields, he said.

If the LIFE Initiative is successful, it could edge out the two other approaches that biomedical scientists are pursuing in order to reduce dependence on donated human organs for transplantation: xenotransplants, or the use of organs from non-human species; and mechanical devices, such as implantable artificial hearts.

Some researchers are worried that xenotransplants could infect humans with new diseases, while engineers have had a difficult time developing an artificial heart that would be small enough and yet reliable enough to be practical.

An External Power Supply

One advantage that a bioengineered heart would have over a mechanical heart is that the bioengineered heart would draw its nutrition from the person's blood, just as an ordinary human heart does.

"It's very hard to compete with that," conceded Alan J. Snyder, an associate professor of surgery and bioengineering at Pennsylvania State University's Hershey Medical Center. He and other researchers there are preparing an artificial heart for tests in humans. Like other artificial hearts under development—and unlike a bioengineered heart—the Penn State heart requires an external power supply (*The Chronicle*, October 2). The bioengineered heart might also be better able to adapt itself in response to changes in its recipient, such as aging; the mechanical hearts such as Penn State's have little ability to adapt.

... the bioengineered heart would draw its nutrition from the person's blood, just as an ordinary human heart does.

Indeed, given the difficulty of the task, even some of Mr. Sefton's colleagues caution against overoptimism. Robert M. Nerem, a professor of engineering in medicine at the Georgia Institute of Technology and a participant in the LIFE Initiative, said that developing a functioning bioengineered heart may take 20 years rather than 10.

Within a decade, he said, it is more realistic to expect the program to develop bioengineered versions of components of organs, rather than full organs, such as bioengineered blood vessels or heart valves.

Developing bioengineered versions of the heart's components would be a highly valuable accomplishment even if the researchers never manage to combine them into a single, working organ, said Mr. Watson of the heart, lung, and blood institute.

For example, mechanical heart valves that are implanted into humans wear out over time and eventually need to be replaced. A bioengineered valve might last longer, he said.

Mr. Sefton said that one intermediate product of the research might be development of a "myocardial patch," or a piece of bioengineered heart tissue that could be stitched into a damaged heart to repair it.

"It could be a very, very useful thing, even if you only got part way" toward a full organ, agreed M.I.T.'s Mr. Langer.

But the goal remains creating full, bioengineered replacements for human organs, Mr. Sefton insisted. "I take exception to the word 'far-fetched,'" he told the meeting.

He prefers another description. "It's a long-term project. There's a lot to be done."

From Science Fiction to Science: "The Whole Body Transplant"[2]

BY MALCOLM W. BROWNE
NEW YORK TIMES, MAY 5, 1998

To lend credibility to his fraudulent account of an execution, the pretentious Pooh-Bah in Gilbert and Sullivan's *The Mikado* sings:

Now though you'd have said that head was dead
(For its owner dead was he),
It stood on its neck, with a smile well-bred,
And bowed three times to me!

For a century audiences have chuckled over Gilbert's whimsy, but last week, viewers of ABC's evening news watched something reminiscent of Pooh-Bah's story that was no laughing matter: a rhesus monkey's severed head, connected by tubes and sutures to the trunk of another monkey, and showing unmistakable signs of consciousness.

The demonstration was an unsettling reminder that an organism, human included, is the sum of many mechanical and chemical parts that ordinarily work in concert but can be made to survive as disembodied entities.

Some viewers certainly flinched while watching the program, but it is worth remembering that modern surgery began with the kind of body-parts experiments that inspired Mary Wollstonecraft Shelley to write *Frankenstein*. Three decades before the publication of her novel, the abundance of severed heads from the guillotine of the French Revolution had kindled a new field of study for some surgeons and scholars.

In the televised presentation last week of his handiwork, Dr. Robert J. White, a 72-year-old professor of neurosurgery at the Case Western Reserve University School of Medicine in Cleveland, demonstrated the grafting of the trunk of one monkey to the head of another in what he calls a "whole body transplant."

Most of the experiments on monkeys, cats and dogs described in the broadcast were first conducted in the 1960's, but Dr. White believes the time is ripe for similar body transplants on humans. He acknowledged in an interview with *The New York Times* that reconnecting the millions of neurons bridging the brain with the spinal column is as yet impossible, and that a person (or rather, a head) who acquired a new body in this way would be paralyzed and insensible from the neck down. But the brain would retain its memory, its intellect, its perception of sight and sound, and its sense of self.

"For a quadriplegic who is already paralyzed, the main cause of death is generally the eventual failure of several organs," Dr. White said. "If such a person were to be given a new body, it would be a new lease on life, even though he or she would still be paralyzed."

The transplanted heads of monkeys evince little tolerance of their executioners. "You now have, as these animals showed, total capability of seeing, hearing and tasting. And if you get your finger too near the mouth of one of these animals, it will bite it off," Dr. White said.

But what is it like to be a severed head?

"I happen to believe that what you and I are is basically within the 3 1/2 pounds of tissue between our ears," Dr. White said. "I think the mind and soul are within the brain. I expressed that view to the Holy Father once, but I don't believe he was convinced."

People have always considered the head to be a special and somewhat mysterious body part. From the trophy hunters of the Amazon headwaters to the followers of Britain's King Charles I (who reverentially sewed back his severed head after he died on the block), people have held the head in special regard.

But the "Age of Enlightenment" in France, which brought the guillotine into prominence, opened a new field of study. As the revolution claimed a rich harvest of heads, surgeons began testing pet theories with the help of electrostatic generators and other inventions of the late 18th century.

The new investigatory spirit was inspired by thinkers like Voltaire and natural philosophers like Antoine Lavoisier, who is remembered today as one of the founders of modern chemistry. (Lavoisier lost his own head to the guillotine—a device promoted by and named for a medical doctor, Joseph Ignace Guillotin. He considered decapitation to be a humane alternative to hanging, disemboweling and other popular 18th-century modes of execution.)

But from the very beginning of France's Reign of Terror (1793-94), some doubted whether death by beheading was instantaneous. Gallows anecdotes recounted in a book by the French historian Andre Soubiran told of heads with moving lips, blinking eyes and other signs of life; witnesses claimed that the lips of some severed heads formed the shapes they would take if they were capable of screaming. Surgeons proposed that condemned prisoners be persuaded to try to demonstrate post-beheading consciousness by some agreed-upon facial signal. (There are no records that this stratagem ever succeeded.)

Dr. White and other experts say the brain remains alive for at most a few seconds after decapitation, until the lack of circulating blood starves it of oxygen and the nutrients needed to continue metabolism. Is the brain conscious during those last few seconds? Dr. White doubts it.

France abolished capital punishment in 1981, but the subsequent lack of fresh human heads has not impeded progress in the transplantation of other body organs; the replacement of kidneys, livers and hearts has become almost routine.

The origins of such lifesaving surgery are grounded in a long history of gruesome experiments.

Among the most famous of the body-parts experimenters in modern times was Dr. Alexis Carrel, a French-American surgeon who was awarded the 1912 Nobel Prize in Medicine for devising the technique of suturing blood vessels together. In later years he and Charles A. Lindbergh, the famous aviator, collaborated in building a glass heart and in keeping organs removed from animals alive for long periods.

In the 1930's Carrel and Lindbergh championed political views compatible with some of those of the Nazis, and in World War II they lost much of their popularity in the United States. But Carrel is nevertheless remembered for his invention of techniques that have made organ transplantation possible.

Perhaps human suffering will one day be reduced by the suffering of a legion of wretched animals who have lost their heads to science. The balance sheet has yet to be made out.

Animal Transplants: Safe or Sorry?[3]

BY AMY OTCHET
UNESCO COURIER, MARCH 2000

Just over 15 years ago, Baby Fae became a household name when American doctors replaced her ailing heart with the heart of a baboon. From Australia to Brazil, millions were riveted by the news reports and followed her progress with baited breath. If they were half-expecting a Frankenstein's monster, they were disappointed. Baby Fae looked like any other wee newborn child. But the sci-fi lullaby turned grim 20 days after the transplant. On November 15, 1984, Baby Fae died. Later her mother angrily maintained that she hadn't been informed about the possible dangers involved. Unbeknownst to the doctors at the time, however, the stakes concerned far more than a single infant's life. We now know there is a possibility that the transplantation of animal organs into humans might unleash infectious diseases similar to AIDS.

The scientific community is only beginning to understand the possible risks of xenotransplantation—the use of animal organs and tissue in "spare-part" surgery for humans. Over the past century, there have been about 25 documented cases of such organ transplants, the most recent case being reported in 1993. Kidneys, hearts and livers from baboons and monkeys were the organs of choice. Survival rates were dismal—most patients died within weeks. Today, however, thanks to progress in biotechnology and drug treatments, there is renewed interest in opening another round of human experiments. Scientists in the United States have already begun implanting pig cells to treat patients with Parkinson's disease and diabetes. Others await the green light to begin transplanting organs from pigs into people.

This growing interest is matched by rising fear, however, that an animal virus could jump from the pig to the human patient, spread to others and unleash a pandemic. When viruses cross the species barrier, the results can be catastrophic. At least one strain of HIV is believed to have jumped from monkeys to people after a single infectious event 60 years ago. The influenza epidemic of 1918-19, which killed tens of millions, may have been triggered by a pig infecting one person.

---

3. Article by Amy Otchet from the *UNESCO Courier* March 2000. Copyright © *UNESCO Courier.* Reprinted with permission.

Xenotransplantation thus confronts the whole of society—not just individual patients—with the promise of saving thousands of lives and the possible risk of causing tremendous harm. The lack of scientific data transforms the safety issues into an ethical dilemma which scientists alone cannot answer.

Human transplantation has been a victim of its own success. Surgeons can now transplant about 25 different kinds of human organs and tissue, and survival rates are constantly improving (60

Xenotransplantation . . . confronts the whole of society—not just individual patients—with the promise of saving thousands of lives and the possible risk of causing tremendous harm.

---

per cent of patients live more than five years). More than one million people worldwide have benefited since 1954, when the first transplant was made. But supply cannot meet demand. In the United States, for example, 3,900 people died while waiting for an organ transplant in 1996, compared to around 1,500 in 1988.

There are also strong economic arguments in favour of xenotransplantation. About 700,000 patients suffering from kidney disease worldwide are strapped to dialysis machines at an annual cost of about $19 billion, according to the Organization for Economic Co-operation and Development (OECD). It costs about 60 per cent less to transplant a kidney than to keep the patient on life-long dialysis. Success in xenotransplantation could open up an international market worth $6 billion plus another $5 billion in related drug treatments (to prevent the immune system from rejecting the organs). One of the biggest contenders is the pharmaceutical giant Novartis—which not only produced cyclosporin A, the leading drug used in human transplants, but also owns Imutran, a UK-based biotech company famous for genetically engineering pigs for xenotransplantation.

Pigs are the xeno-prize-winners. Nonhuman primates, like baboons, have been ruled out because their biological similarities to human beings could increase the risk of disease transmission, so brutally highlighted by the AIDS and Ebola viruses. Many people also have ethical qualms about using our "cousins" for spare parts, whereas we have been slaughtering and eating pigs for many centuries. Finally, pigs are easier to breed and genetically engineer.

Duping the Human Immune System

The human body would normally consider a pig organ to be a dangerous "foreign agent" and kill it within minutes by cutting off its blood supply. Imutran and other laboratories are trying to get past this defence (immune) system by lining the pigs' organs with human proteins through genetic engineering. These proteins endow the pig organ with a kind of human disguise. After a time, however, the human body would gradually realize that the pig organ isn't acceptable and would launch an attack on it. Imutran is now trying to develop new drugs and may add more human genes to the pigs, says Dr. Corinne Savill, the company's chief operating officer.

These "designer pigs" may, however, make it easier for the animals' germs to infect people, says Dr. Robin Weiss of University College London. About two years ago, Weiss began publishing articles showing how pig viruses could hide behind the human proteins (added to the pigs) and slip past a patient's immune system. The human proteins might also invite the viruses into human cells. For example, one of the human proteins used by Imutran and other biotech companies is CD55. This protein makes the human body vulnerable to several polio-related viruses. Suppose, says Weiss, that pigs have similar viruses. Ordinarily, they wouldn't affect humans because of genetic differences. But imagine that the pig viruses learn (through genetic modification) to use CD55 and infect the patient who receives the pig organ. Once a pig disease has crossed over into a single human being, it could mutate further and spread to others.

None of these questions would matter if the pigs could be issued with a "clean bill of health," says Weiss. Imutran, for example, is trying to breed germ-free pigs in tightly controlled facilities. But even in a hermetically sealed tank all the risks would not be eliminated, particularly those arising from viruses known as PERV retroviruses which are found in the animal's genes. Weiss has focused on three strains of PERV, which have been described as "second cousins" to HIV or human immunodeficiency virus which causes AIDS. Two of them can infect human cells.

Weiss's findings sent shockwaves through public health organizations and the industry. Researchers immediately set out to survey as many patients as they could find who had been exposed to porcine tissue. Out of about 175 patients screened, none were infected by PERV.

"It's a relief," says Weiss, but it's not conclusive. "This particular virus is not going to be highly contagious but this doesn't necessarily mean that xenotransplantation is safe." The patients in question received porcine tissue not organs, whose sheer volume may increase the risk of infection. Second, they were screened for the three known retroviruses. What about unknown germs? Weiss also wonders whether the retroviruses might be hiding somewhere in the body and become more powerful over time.

François Meslin of the World Health Organization describes a worst-case scenario: a xeno-patient harbours an undetected virus which is passed to others by sexual intercourse, a likely means of transmission. As the virus moves from one host to another, it becomes increasingly dangerous.

"You can take a lot of precautions but you never know how far to go to bring the risks down to an acceptable level," says Meslin, who points to the case of bovine spongiform encephalopathy—"mad cow disease"—as the closest example of this kind of dilemma. Since Weiss sounded the alert, public health authorities around the world have observed a de facto moratorium on human experiments with xenotransplantation, which doctors describe as clinical trials. This doesn't mean that they are giving up on the research, however. What is happening is that most Western governments are setting up special advisory or regulatory bodies to review any future clinical trials and prepare strict guidelines for monitoring them.

"You can take a lot of precautions but you never know how far to go to bring the risks down to an acceptable level." — **François Meslin, World Health Organization**

The U.S. and Britain, who are the leaders in xeno-research, are currently finalizing guidelines to monitor not just the patients, but their family-members and health-workers. While the authorities refuse to release the details, it has been said that xeno-patients should refrain from having children, marrying or even travelling internationally. This sounds like a replay of discussions concerning people infected with HIV, says Prof. Bartha Maria Knoppers, a Canadian bioethicist regularly called on by OECD to examine the ethics of xenoresearch. Close monitoring of patients should not mean trampling on their human rights, Knoppers believes. "Besides," she says, "are we really going to be able to enforce these conditions?" Would it be possible, for example, to take legal steps to prevent a patient from deciding to have a child two years after a clinical trial?

Assessing the Risk of Viral Infections

The sponsors of these trials—mostly biotech companies—will also come under the microscope. Ordinarily, proposals to test new medical procedures or treatments on humans are considered commercial secrets which are restricted to the company involved and the relevant governmental regulatory body, e.g. the U.S. Food and Drug Administration (FDA) which is responsible for approving tests on humans for new medical drugs or procedures. This is not so with xenotransplantation, however. In the United States, a special advisory committee composed of some 15 scientific experts will openly review all requests to test animal organs in people before the FDA renders a decision.

"Absence of evidence is not absence of risk," says Phil Noguchi, director of the FDA's division of cellular and gene therapies. For example, two years ago bio-tech companies proudly insisted that their pigs were germ-free. But when news of the PERV retroviruses came out, Noguchi says, the companies suddenly "worked a lot harder" to examine the risks. Noguchi also maintains that viral infections are difficult to evaluate. "They may not happen often enough to be a publishable event," he says. "But in a clinical experiment one tenth of a sixth of a chance of infection is a whole lot." This is why the FDA is counting on help from the advisory committee to look in all the "nooks and crannies" where a virus might hide. "We're still in a very difficult position because to a large extent we rely on industry to provide the proof that a clinical experiment is safe," Noguchi notes. "But we also depend on our own scientists."

For Noguchi, the advisory body also scores extra points by offering a "public forum" to discuss the ethical issues—from animal welfare concerns to deciding who should receive an animal organ. First, companies might focus on patients with the greatest chances of surviving instead of those most in need of a transplant. Second, it will take years—if ever—before pig organs are effective. For a patient on the verge of death a pig organ could buy a few extra weeks until a human organ becomes available. Should such patients receive priority on the waiting list? There is a third major concern, says John Davies of the U.S. National Kidney Foundation, the world's largest non-profit charity for kidney patients. "We don't want people to stop donating the organs of their loved ones [for human transplants] after the first animal transplant trials," says Davies, "because they think the problem has been resolved." While Davies supports xeno-research, he is not convinced that it will prove successful.

The problem is that these discussions take place between people with vested interests of one kind or another. They fail to include the largest group affected, the public. "It can be argued that there should be some sort of 'community consent,'" says Dr. A. S. Daar of Oman, who chaired the World Health Organization's consultation on xenotransplantation. "Is the FDA mandate to protect the public enough of a proxy for community consent?" asks Daar. "Until the public is informed about the issues and is discussing them, I don't think you would want to go ahead with clinical trials."

Daar is looking for ways of stimulating public debate and consent by working with an international committee of "concerned" citizens—mostly scientists and bioethicists brought together by the prestigious Hastings Center, a bioethics think-tank in New York. The aim is to help countries to develop national but non-governmental committees of individuals from various walks of life—

"Until the public is informed about the issues and is discussing them, I don't think you would want to go ahead with clinical trials."—**Dr. A. S. Daar, a WHO consultant on xenotransplantation**

---

economics, law, religion, the media, etc.—who will take time to become informed about xenotransplantation but don't themselves have any vested interests in the matter. These committees would hold "consensus conferences"—what Americans call "townhall meetings"—to spread information about xenotransplantation and gauge the public's response to it.

This idea is the brainchild of Fritz Bach, a scientist at Harvard Medical School who first called for a legal moratorium on clinical trials in the U.S. Bach is often portrayed as a virulent opponent of xenotransplantation and yet he is a leading scientist in the field as well as a paid consultant for Novartis. His corporate sponsor isn't thrilled by his idea of townhall meetings. "You cannot go forward by publicly discussing such complicated subjects," says Imutran/Novartis's Dr. Savill. "Now, whether governments have set up the right agencies is another question. Like anything else in society, are the right people in the right place making the decisions? And does the public have confidence in them?"

Bach isn't convinced. "Some insiders think this [moratorium and public consultation] would be a good thing for Novartis," he says. "Just look at the tremendous hullabaloo over genetically modified organisms. If a blue ribbon committee—without any connections to Monsanto—had informed the public that they thought the [GM] food was safe on the shelves, we wouldn't have had this reaction. The scariest thing is always the unknown."

Mind over Matter[4]

Brain-driven prostheses move from science fiction to science.

BY DAMARIS CHRISTENSEN
SCIENCE NEWS, AUGUST 28, 1999

Blinking is neither the fastest nor the most accurate way to communicate. It was, however, the only way left for Jean-Dominique Bauby, editor-in-chief of France's ELLE magazine. He suffered a stroke in 1995 that left him almost totally paralyzed, yet he was determined to write a book. Bauby dictated his story by having a colleague recite the alphabet to him and then blinking his left eyelid to select each letter of each sentence.

"Something like a giant invisible diving bell holds my whole body prisoner," he wrote in *The Diving Bell and the Butterfly* (Knopf), published just before he died in 1997. In the book, he recorded the difficulties of communicating with his friends, family, and doctors.

As many as a million people in the United States are locked inside their bodies the way Bauby was. They retain full control of their minds but can't breathe, eat, or move on their own because of injury or disease. Several researchers are looking for ways for these so-called locked-in patients to communicate.

The use of technology to overcome disabilities spans human history, ranging from simple crutches to modern prosthetic arms and hands that can move and grip with remarkable precision. However, these external devices only go so far. For years, science fiction writers and some scientists have dreamed of compensating for damage by connecting a person's brain directly to a prosthetic device or a computer running one. Bauby's task would have been easier if such a brain-to-computer link had been available to him.

Despite the appeal of such connections, hooking into the brain is no easy task. Even the simplest of everyday movements requires complex computations. The brain is constantly making calculations and sending out signals to hundreds of muscles in healthy arms, hands, legs, and feet.

Researchers have long thought that it should be possible to tap into the electric signals produced by nerve cells, or neurons, and use them to control the shifting path of a cursor on a computer screen, the movement of a wheelchair, or the grasp of a robotic arm. The problem is that no one yet fully understands the complex electrical signals the brain sends, says Eberhard E. Fetz of the University of Washington School of Medicine in Seattle. In addition, neural signals have turned out to be far from easy to intercept.

The advances in miniaturization that have made laptop computers commonplace, however, have also allowed scientists to eavesdrop more accurately on the brain. While complex brain-to-computer interfaces still lie many years in the future, a few recent studies have made direct neural control of computers or prosthetic devices appear more promising.

Two groups have worked with locked-in patients, helping a few to communicate with the world by moving a cursor on a computer screen. Another group has trained rats to move a lever not by using their muscles but by producing certain kinds of nerve signals in their brains.

"Extracting signals directly from the brain to directly control robotic devices has been a science fiction theme that seems destined to become fact," Fetz says.

One approach to robotic control is to tap into electrical noise generated by the brain's normal activity. Electrodes on the scalp can measure the tiny amounts of current generated by nerve cells in the brain as they fire. People can use biofeedback techniques to learn to control the patterns of these electroencephalograms, or EEGs. Because it does not require surgery, this approach is considered safe. Learning to control the patterns is time-consuming, however. Moreover, some researchers say that EEGs may not contain enough information to enable patients to quickly and gracefully manipulate an object in several dimensions.

> *"Extracting signals directly from the brain to directly control robotic devices has been a science fiction theme that seems destined to become fact."*—Eberhard E. Fetz, University of Washington School of Medicine, Seattle

A German team reported in the March 25 *Nature* that two locked-in patients have learned to operate a spelling device by controlling their EEG brain responses. Both patients suffer from advanced amyotrophic lateral sclerosis, or ALS, and can't breathe or eat on their own.

The German device flashes half of the alphabet up on each side of a computer screen. The ALS sufferer selects one half or the other by controlling his or her EEG signals. The patient thus repeatedly divides the alphabet until a single letter is chosen.

Writing in this way is not much faster than Bauby's painstaking technique—a person can select about two letters each minute—but the system is accurate and allows these people to communicate on their own, says research leader Niels Birbaumer of the University of Tüüebingen in Germany. "Even a slow spelling device is helpful," he says.

In similar work, Jonathan R. Wolpaw at the State University of New York at Albany has shown that normal and paralyzed volunteers can use other types of EEGs to control a cursor on a computer.

In another approach, scientists implant electrodes directly into a person's brain to detect signals from neurons in the area that once controlled an arm. A research team based at Emory University in Atlanta has inserted tiny electrodes, each surrounded by a glass cone, into the brains of three locked-in patients. The cones contain proteins that encourage nerve cells to grow near the electrode. Bursts of activity detected by the electrodes can drive a cursor across a computer screen. Each electrode may measure the activity of a few nerve cells. So far, the researchers have implanted just one or two electrodes into each patient.

One patient learned to control a cursor but died of ALS just 2 months after Bakay implanted the electrodes.

Another patient, named J. R., uses the cursor to select different icons on a screen. Each conveys a different message—for example, that he's thirsty. The researchers have also copied a computer keyboard onto the screen so he can slowly type out messages.

The technology has its limits. Currently, J. R. can control the cursor for just under an hour each day. It's hard for him to maintain the focus required to move the cursor, and he is very sick and tires easily, says Roy A. E. Bakay of Emory.

"This is a lot of work for him, but this is one of the few ways he can get messages out," he says.

A small but promising study in rats uses tiny electrode arrays—each electrode in contact with a different neuron. The results suggest that animals can directly control a robotic device with their brain activity. Researchers in Philadelphia and Durham, N.C., implanted the arrays in the motor cortexes of half a dozen rats. This is the area of the brain involved in movement.

The researchers measured the brain activity of the normal, healthy rats as they learned to push a lever attached to a robotic arm. If the animals pushed the lever hard enough, the robotic arm carried water to a place where the animal could drink. The researchers then fed the rats' brain signals to a computer, which identified a burst of electrical activity released just before an animal pushed the lever. Each rat produced similar but not identical signals. The researchers then devised a program to move the robotic arm as soon as a rat's brain made these electrical signals. Some rats learned over time that they didn't actually have to push the lever to get the robotic arm to move, says research leader John K. Chapin of the MCP Hahnemann School of Medicine in Philadelphia.

"Previously, researchers have focused on single neurons in the motor systems. We took a broader look," he says. They found that a larger sample of the many neurons involved in movement eased the task of finding reliable, detectable signals to trigger a prosthetic device.

Although many technical hurdles remain, Chapin says, "we believe we have all the key elements to be able to make this technology one that could, in the not-so-distant future, make a substantial difference in the lives of people who are limited in their physical abilities but not their neurologic capabilities."

This is the first study showing that simultaneous recordings from a number of neurons can immediately trigger movement of an external device, Fetz says. "It's surprising that [in rats] neural activity could be dissociated from movement," he says.

Because of their safety, devices that externally measure EEGs may initially be more widely used than electrodes implanted in the brain, says William J. Heetderks of the National Institute of Neurological Disorders and Stroke in Bethesda, Md. Despite Bakay's success, widespread use of implanted electrodes in humans is probably at least a decade away, Heetderks says.

There is a large leap from showing that rats can move a robotic arm back and forth to demonstrating that humans can continuously operate a prosthetic device to mimic a human arm, or even can move a wheelchair, Fetz says.

The tiny electrodes used by Bakay and Chapin stem from a quiet revolution in miniaturization of bioelectronic components. "We don't yet know what the limitations of this new, emergent technology are" says Richard A. Normann of the University of Utah in Salt Lake City.

"Chapin's work is not enough to prove that arrays of electrodes implanted in the brain can control a robotic arm in a graded, proportional fashion," he says. "One of the remarkable things about humans and animals is that they can move so gracefully, and we should strive for that."

Bakay and his colleagues say that their lab's electrodes are sensitive enough to measure subtle differences in the rate at which nerve cells fire. If so, the devices might work as dimmers rather than simple on-off switches. Such proportioned control is critical if paralyzed patients are to accurately tilt a hospital bed, run a wheelchair, or move a prosthetic arm, he says.

"Right now, you can walk your way through the world without thinking consciously about what you are doing," says Normann. "It would be nice for a completely paralyzed person to likewise have multiple degrees of control over their movement—for example, to be able to make a wheelchair go faster and turn left all at the same time."

Many challenges lie ahead. For humans to be able to produce complex external movements, researchers will have to establish links with many more neurons than they have in the experiments so far. That task will require smaller, longer-lasting electrodes, and surgeons will have to place the devices more precisely. To register brain signals over a long period, electrodes must not move, cause scarring, or repel growing nerve cells, says Heetderks.

> *"It would be nice for a completely paralyzed person to . . . have multiple degrees of control over their movement . . . ."*—
> **Richard A. Normann, University of Utah, Salt Lake City**

Researchers may have to look beyond the primary motor cortex to measure enough neurons to control complex movement, notes Chapin.

One possible problem with using signals from the primary motor cortex emerges from some studies showing that the brain reconfigures itself once sensory input has changed, says Miguel A. L. Nicolelis of Duke University Medical Center in Durham, who worked with Chapin. After an injury, areas of the brain that used to control the limb that became paralyzed may degenerate or rewire themselves to control other parts of the body. Such rewiring might mean that the implanted electrodes are not measuring electrical activity from neurons that would be involved with movement of the limb. The changes may make it less likely that patients can operate a prosthetic device by simply thinking about moving a missing or paralyzed limb.

Nicolelis hopes to address the possible rewiring of the brain as part of his current study of owl monkeys. Because monkeys have bigger brains than rats, however, it may be harder to identify neurons involved in a particular kind of movement, he says.

Richard A. Andersen of the California Institute of Technology in Pasadena has a possible solution to the motor cortex's degeneration or rewiring. In new experiments, he plans to target a different area of the brain called the posterior parietal cortex. This part of the brain seems to be responsible for taking in sensory information and for planning movements, he says.

Andersen and his colleagues have implanted in monkeys electrodes similar to those used by Chapin's group. They hope to train the monkeys to move images of prosthetic arms on a computer screen. From there, it's just a small step to actual movement of a robotic prosthetic device, says Andersen.

"The eventual hope is to use these technologies to assist patients who are paralyzed, but the long-term safety and reliability of the devices must be proven first," Heetderks says. "When you get up in the morning, you expect your arm to work the same way it did yesterday."

Another issue for researchers is portability of any EEG or neuron-monitoring devices. All strategies for monitoring brain activity, so far, require tethering patients to equipment.

Despite the challenges that remain, researchers hope that the phrase mind over matter will eventually become more than a cliché. "Right now, we're just taking a few baby steps," says Bakay. "I don't think anybody yet knows the best way to do this, and maybe we will eventually use aspects of all this work."

Even imperfect steps offer great benefit to severely paralyzed patients, the researchers point out. Birbaumer's first patient spent 16 hours writing a thank-you note to him, one letter at a time.

Cyborg 1.0[5]

Kevin Warwick outlines his plan to become one with his computer.

By Kevin Warwick
Wired, February 2000

I was born human. But this was an accident of fate—a condition merely of time and place. I believe it's something we have the power to change. I will tell you why.

In August 1998, a silicon chip was implanted in my arm, allowing a computer to monitor me as I moved through the halls and offices of the Department of Cybernetics at the University of Reading, just west of London, where I've been a professor since 1988. My implant communicated via radio waves with a network of antennas throughout the department that in turn transmitted the signals to a computer programmed to respond to my actions. At the main entrance, a voice box operated by the computer said "Hello" when I entered; the computer detected my progress through the building, opening the door to my lab for me as I approached it and switching on the lights. For the nine days the implant was in place, I performed seemingly magical acts simply by walking in a particular direction. The aim of this experiment was to determine whether information could be transmitted to and from an implant. Not only did we succeed, but the trial demonstrated how the principles behind cybernetics could perform in real-life applications.

Eighteen months from now, or possibly sooner, I will conduct a follow-up experiment with a new implant that will send signals back and forth between my nervous system and a computer. I don't know how I will react to unfamiliar signals transmitted to my brain, since nothing quite like this has ever before been attempted. But if this test succeeds, with no complications, then we'll go ahead with the placement of a similar implant in my wife, Irena.

My research team is made up of 20 scientists, including two who work directly with me: Professor Brian Andrews, a neural-prosthesis specialist who recently joined our project from the University of Alberta in Canada, and professor William Harwin, a cybernetics expert and former codirector of the Rehabilitation Robotics Labora-

5. Article by Keven Warwick from *Wired Magazine* February 2000. Copyright © *Wired Magazine*. Reprinted with permission.

tory at the University of Delaware in the US. The others are a mixture of faculty and researchers, divided into three teams charged with developing intelligent networks, robotics and sensors, and biomedical signal processing—i.e., creating software to read the signals the implant receives from my nervous system and to condition that data for retransmission.

We are in discussions with Dr. Ali Jamous, a neurosurgeon at Stoke Mandeville Hospital in nearby Aylesbury, to insert my next implant, although we're still sorting out the final details. Ordinarily, there might be a problem getting a doctor to consider this type of surgery, but my department has a long-standing research link with the hospital, whose spinal-injuries unit does a lot of advanced work in neurosurgery. We've collaborated on a number of projects to help people overcome disabilities through technical aids: an electric platform for children who use wheelchairs, a walking frame for people with spinal injuries, and a self-navigating wheelchair. While Jamous has his own research agenda, we are settling on a middle ground that will satisfy both parties' scientific goals.

My first implant was inserted by Dr. George Boulos at Tilehurst Surgery in Reading into the upper inside of my left arm, beneath the inner layer of skin and on top of the muscle. The next device will be connected to the nerve fibers in my left arm, positioned about halfway between my elbow and shoulder. (It doesn't matter which arm carries the implant; I chose my left because I'm right-handed, and I hope I will suffer less manual impairment if any problems arise during the experiment.) Most of the nerves in this part of the body are connected to the hand, and send and receive the electronic impulses that control dexterity, feeling, even emotions. A lot of these signals are traveling here at any given time: This nerve center carries more information than any other part of the anatomy, aside from the spine and the head (in the optic and auditory nerves), and so is large and quite strong. Moreover, very few of the nerves branch off to muscles and other parts of the upper arm—it's like a freeway with only a few on- and off-ramps, providing a cleaner pathway to the nervous system.

While we ultimately may need to place implants nearer to the brain—into the spinal cord or onto the optic nerve, where there is a more powerful setup for transmitting and receiving specific complex sensory signals—the arm is an ideal halfway point.

This implant, like the first, will be encased in a glass tube. We chose glass because it's fairly inert and won't become toxic or block radio signals. There is an outside chance that the glass will break, which could cause serious internal injuries or prove fatal, but our previous experiment showed glass to be pretty rugged, even when it's frequently jolted or struck.

One end of the glass tube contains the power supply—a copper coil energized by radio waves to produce an electric current. In the other end, three mini printed circuit boards will transmit and receive signals. The implant will connect to my body through a band that wraps around the nerve fibers—it looks like a little vicar's collar—and is linked by a very thin wire to the glass capsule.

The chips in the implant will receive signals from the collar and send them to a computer instantaneously. For example, when I move a finger, an electronic signal travels from my brain to activate the muscles and tendons that operate my hand. The collar will pick up that signal en route. Nerve impulses will still reach the finger, but we will tap into them just as though we were listening in on a telephone line. The signal from the implant will be analog, so we'll have to convert it to digital in order to store it in the computer. But then we will be able to manipulate it and send it back to my implant.

No processing will be done inside the implant. Rather, it will only send and receive signals, much like a telephone handset sends and receives sound waves. It's true that onboard power would increase our options for programming more complex tasks into the implant, but that would require a much larger device. While a 1-inch-long glass tube isn't obtrusive, I really don't fancy an object the size of an orange built into my arm.

We'll tap into my nerve fibers and try a progression of experiments once my new implant is switched on. One of the first will be to record and identify signals associated with motion. When I waggle my left index finger, it will send a corresponding signal via the implant to the computer, where it will be recorded and stored. Next, we can transmit this signal to the implant, hoping to generate an action similar to the original. I will consider the test a fantastic success if we can record a movement, then reproduce it when we send the signals back to the arm.

Pain also provides a distinctly clear electronic signal on the nervous system as it moves from its point of origin to the brain. We intend to find out what happens if that signal is transmitted to the computer and then played back again. Will I feel the same sensation, or something more akin to the phantom pains amputees "feel" in their missing limbs? Our brains associate an ache with a specific point on the body; it will also be interesting to see whether this sensation can be manipulated by slightly modifying the signal in the computer and then trying to send it to another area.

When the new chip is in place, we will tap into my nerve fibers and try out a whole new range of senses.

We will then attempt this exercise with emotional signals. When I'm happy, we'll record that signal. Then, if my mood changes the next day, we'll play the happy signal back and see what happens.

I am most curious to find out whether implants could open up a whole new range of senses. For example, we can't normally process signals like ultraviolet, X rays, or ultrasound. Infrared detects visible heat given off by a warm body, though our eyes can't see light in this part of the spectrum. But what if we fed infrared signals into the nervous system, bypassing the eyes? Would I be able to learn how to perceive them? Would I feel or even "see" the warmth? Or would my brain simply be unable to cope? We don't have any idea—yet.

The potential for medical breakthroughs in existing disabilities is phenomenally important. Might it be possible to add an extra route for more senses or to provide alternative pathways for blind or deaf people to "see" or "hear" with ultrasonic and infrared wavelengths? Perhaps a blind person could navigate around objects with ultrasonic radar, much the way bats do. Robots have been programmed to perform this action already, and neuroscientists have not dismissed the idea for humans. But few people have ever had their nervous systems linked to a computer, so the concept of sensing the world around us using more than our natural abilities is still science fiction. I'm hoping to change that.

People have asked me, too, whether it would be possible to get high from drugs, store those signals, and then return them to the nervous system later to reproduce the sensation. To that end, I plan to have a glass or two of wine and record my body's reaction, captured in exactly the same way I "saved" movement or pain. The following day, I will play back the recorded signals. As my brain tries to make sense of these, it might search for past experiences, trying to put things in terms of what it already knows. Thus, when my brain receives the "drunk" signal, it might believe it is indeed intoxicated. Varying on that theme, perhaps particular electronic patterns can be transmitted to the nervous system to bring about a sensation equivalent to that of drinking bourbon or rum.

If this type of experiment works, I can foresee researchers learning to send antidepressant stimulation or even contraception or vaccines in a similar manner. We have the potential to alter the whole face of medicine, to abandon the concept of feeding people chemical treatments and cures and instead achieve the desired results electronically. Cyberdrugs and cybernarcotics could very well cure cancer, relieve clinical depression, or perhaps even be programmed as a little pick-me-up on a particularly bad day.

We don't know how much the brain can adapt to unfamiliar information coming in through the nerve branches. Our hunch is that the brain of a young child is pliable, so that it might well be able to take in new sensory information in its own right. In response to the additional input, the nerve fibers linked to an implant might begin

---

Cyberdrugs and cybernarcotics could very well cure cancer, relieve clinical depression, or perhaps even be programmed as a little pick-me-up on a particularly bad day

---

to grow thicker and more powerful with the ability to carry more and different kinds of information. A 45-year-old brain like mine is another matter. In the absence of any previous sensory reference, will my brain be able to process signals that don't correspond precisely to sight, sound, smell, taste, or touch? It will probably deal with something like X-ray stimulation in terms of the signals it thinks most similar. Depending on its best guess, I might feel pain, tension, or excitement. But we want to avoid feeding in too much noise, as that could be distinctly risky. I do worry that certain kinds of raw input could make me crazy. For me, in any case, all these experiments are worth doing just to see what might happen. If the results aren't encouraging, then—what the hell—at least I tried.

I plan to keep my next implant in place for a minimum of a week, possibly up to two. If the experiments are successful, we would then place implants into two people at the same time. We'd like to send movement and emotion signals from one person to the other, possibly via the Internet. My wife, Irena, has bravely volunteered to go ahead with his-and-hers implants. The way she puts it is that if anyone is going to jack into my limbic system—to know definitively when I'm feeling happy, depressed, angry, or even sexually aroused—she wants it to be her.

Irena and I will investigate the whole range of emotion and sensation. If I move a hand or finger, then send those signals to Irena, will she make the same movement? I think it likely she'll feel something. Might she feel the same pain as I do? If I sprained my ankle, could I send the signal to Irena to make her feel as though she has injured herself?

We know that different people have varying emotional responses to the same stimulus. If I send a particular signal to her, will she recognize it in the same way? Based on my own reaction to having my emotional impulses replayed on my nervous system, we will have a preliminary idea of what Irena might experience, but we are entering progressively uncharted territory once we attempt to relay prerecorded signals. What her brain can comprehend in terms of my neural impulses is completely unknown. Yet if Irena's brain can make out, even roughly, my incoming signals, then I believe her own stored knowledge will be able to decipher the information into a recognizable sensation or emotion.

We would also like to demonstrate how the signals could be sent over the Internet. One of us will travel to New York, and the other will remain in the UK. Then we'll send real-time movement and emotion signals from person to person across the continents. I am terrified of heights. If I'm staying on the 16th floor of a hotel in the US and I transmit my signals to Irena, how will they affect her? How far could we go in transmitting feelings and desires? I want to find out. What if the other person became sexually aroused? Could we record signals at the height of our arousal, then play these back and relive the experience? (As keen as I am to know the answer here, I have difficulty imagining what the scientific press might make of it.)

Will we evolve into a cyborg community? Linking people via chip implants to superintelligent machines seems a natural progression—creating, in effect, superhumans.

We are not the first group to link computers with the human nervous system via implants. Dr. Ross Davis' team at the Neural Engineering Clinic in Augusta, Maine, has been trying to use the technology to treat patients whose central nervous systems have been damaged or affected by diseases like multiple sclerosis, and has been able to achieve basic controls over, for example, muscle function.

In 1997, a widely publicized project at the University of Tokyo attached some of a cockroach's motor neurons to a microprocessor. Artificial signals sent to the neurons through electrodes were then used to involuntarily propel the cockroach, despite what it might have chosen to do. Also, in an experiment published last summer by John Chapin at the MCP Hahnemann School of Medicine in Philadelphia and Miguel Nicolelis at Duke University, electrodes were implanted into rats' brains and used to transmit signals so that the rats merely had to "think" about pressing a lever in order to receive a treat. Researchers were interested to learn that the signals indicating what the rats were about to do appeared in a different part of the brain than the one usually associated with planning.

And I'm amazed by results from a team at Emory University in Atlanta, which to great international interest has implanted a transmitting device into the brain of a stroke patient. After the motor neurons were linked to silicon, the patient was able to move a cursor on a computer monitor just by thinking about it. That means thought signals were directly transmitted to a computer and used to operate it, albeit in a rudimentary way. The Emory team is looking to gradually extend the range of controls carried out.

As for self-experimentation, physicians and scientists have done this throughout history. During the early '50s, US Air Force colonel John Stapp repeatedly strapped his body to rocket sleds and propelled himself to more than 600 mph before hitting the brakes to stop in less than 2 seconds. The military physician's study of the human body's tolerance for crash forces helped improve automobile, airplane, and spacecraft safety. Although Stapp survived his perilous experiments, he suffered eye damage, a hernia, a concussion, and broken bones and permanently impaired his sense of balance.

> *I have been involved with technology all my life, and now I will be able to take my research one step further.*

In 1984, Barry Marshall, a resident at Royal Perth Hospital in Australia, swallowed an ulcer-causing bacteria to show that the organism, and not stress, caused the abdominal ailment. Then there was Werner Forssmann, a German physician so obsessed with learning the intricacies of the human heart that in 1929 he inserted a catheter into an artery in his arm and snaked it all the way to his right auricle. In 1892, another German doctor, Max von Pettenkofer, drank a culture of the bacterium that causes cholera to show that environmental factors must also be present before the germ produces the disease. He was sick for about a week but lived—pure luck, of course, since we now know his hypothesis was erroneous. And Isaac Newton stuck needles into his eyes—for what reason, I'm not entirely sure.

As for me, I am not a foolish scientist putting my life in harm's way. In fact, my next implant will be the culmination of my professional work: working for British Telecom, studying computer engineering and robotics, and teaching the principles of cybernetics. I have been involved with technology all my life, and now I will be able to take my research one step further.

Admittedly, I'm putting the neurological and medical aspects of the operation in the hands of the surgeon. I realize the chance of infection is higher with my second implant, since it will touch the nerve bundles. And connecting to the nervous system could also lead to permanent nerve damage, resulting in the loss of feelings or movement, or continual pain. But I am putting aside my fears and accepting my less-than-absolute understanding of the technical and psychological ramifications inherent in our attempt. I want to know.

I believe this desire—this urge to explore—is intrinsically human. My entire team is venturing into the unknown with me in order to bring humans and technology together in a way that has never been attempted. The excitement of looking over the horizon into a new world—the world of cyborgs—far outweighs the risks. Just think: Anything a computer link can help operate or interface with could be controllable via implants: airplanes, locomotives, tractors, machinery, cash registers, bank accounts, spreadsheets, word processing, and intelligent homes. In each case, merely by moving a finger, one could cause such systems to operate. It will, of course, require the requisite programs to be set up, just as keyboard entries are now required. But such programming, along with the implant owner learning a few tricks, will be relatively trivial exercises.

In the future, we won't need to code thoughts into language—we will uniformly send symbols and ideas and concepts without speaking.

Linking up in this way could allow for computer intelligence to be hooked more directly into the brain, allowing humans immediate access to the Internet, enabling phenomenal math capabilities and computer memory. Will you need to learn any math if you can call up a computer merely by your thoughts? Must you remember anything at all when you can access a world Internet memory bank?

I can envision a future when we send signals so that we don't have to speak. Thought communication will place telephones firmly in the history books. Philosophers point to language in humans as being an important part of our culture and who we are. Certainly, language has had everything to do with human development. But language is merely a tool we use to translate our thoughts. In the future, we won't need to code thoughts into language—we will uniformly send symbols and ideas and concepts without speaking. We will probably become less open, more able to control our feelings and emotions—which will also become necessary, since others will more easily be able to access what we're thinking or feeling. We will still fall back on speech in order to communicate with our newborns, however, since it will take a few years before they can safely get implants of their own, but in the future, speech will be what baby talk is today.

Thought-to-thought communication is just one feature of cybernetics that will become vitally important to us as we face the distinct possibility of being superseded by highly intelligent machines. Humans are crazy enough not only to build machines with an overall intelligence greater than our own, but to defer to them and give them power that matters. So how will humans cope, later this century, with machines more intelligent than us? Here, again, I believe cybernetics can help. Linking people via chip implants directly to those machines seems a natural progression, a potential way of harnessing machine intelligence by, essentially, creating superhumans. Otherwise, we're doomed to a future in which intelligent machines rule and humans become second-class citizens. My project explores a middle ground that gives humans a chance to hang in there a bit longer. Right now, we're moving toward a world where machines and humans remain distinct, but instead of just handing everything over to them, I offer a more gradual coevolution with computers.

From a medical point of view, I was pleased when the first implant was taken out, but I was otherwise quite upset—I felt as though a friend had just died.

Yet once a human brain is connected as a node to a machine—a networked brain with other human brains similarly connected—what will it mean to be human? Will we evolve into a new cyborg community? I believe humans will become cyborgs and no longer be stand-alone entities. What we think is possible will change in response to what kinds of abilities the implants afford us. Looking at the world and understanding it in many dimensions, not just three, will put a completely different context on how we—whatever "we" are—think.

I base this on my own experience with my first implant, when I actually became emotionally attached to the computer. It took me only a couple of days to feel like my implant was one with my body. Every day in the building where I work, things switched on or opened up for me—it felt as though the computer and I were working in harmony. As a scientist, I observed that the feelings I had were neither expected nor completely explainable—and certainly not quantifiable. It was a bit like being half of a pair of Siamese twins. The computer and I were not one, but neither were we separate. We each had our own distinct but complementary abilities. To be truthful, Irena started to get rather worried—jealous, perhaps—when I tried to explain these sensations.

With the new implant, I expect this feeling of connectedness to be much stronger, particularly when emotional signals are brought into the equation. From a medical point of view, I was pleased when the first implant was taken out, but I was otherwise quite upset—I felt as though a friend had just died. With the new implant I might find it impossible to let go, despite the potential for long-term problems were I to retain it.

These desires—which draw me closer to the implant—could ultimately influence my own values and what it means to me to be human. Morals and ethics are an outgrowth of the way in which humans interact with each other. Cultures may have diverse ethics, but, regardless, individual liberties and human life are always valued over and above machines. What happens when humans merge with machines? Maybe the machines will then become more important to us than another human life. Those who have become cyborgs will be one step ahead of humans. And just as humans have always valued themselves above other forms of life, it's likely that cyborgs will look down on humans who have yet to "evolve."

Surprisingly, nobody has reacted to my plans by telling me, "That's impossible"—I think because no one really knows what will happen. When I tell others about my work, more often they are aghast, not really comprehending what I'm talking about. But no scientists have told me I shouldn't be playing God or that what I'm doing is unfeasible or too dangerous. Even so, I am certain that after Alexander Graham Bell said, "Mr. Watson, come here, I want you," the cynics asked, "Why didn't you just walk to the next room and speak to him?" At the time, it was difficult to see where it all might lead. Of course, I don't put myself in the same category as people like Bell or Charles Lindbergh or John F. Kennedy—pioneers who were convinced we could do things like land men on the moon. But I've been inspired by these visionaries, these risk takers, each of whom spent his lifetime obsessively pursuing his goals.

Since childhood I've been captivated by the study of robots and cyborgs. Now I'm in a position where I can actually become one. Each morning, I wake up champing at the bit, eager to set alight the 21st century—to change society in ways that have never been attempted, to change how we communicate, how we treat ourselves medically, how we convey emotion to one another, to change what it means to be human, and to buy a little more time for ourselves in the inevitable evolutionary process that technology has accelerated. In the meantime, I feel like screaming when I have to do paperwork or shop or go to sleep—it's stopping me from getting on with what I really want to do. The next implant cannot come soon enough.

VI.

Genetically Modified Organisms

Editor's Introduction

The saying goes "You are what you eat." But in the case of today's consumers, few people actually know what they eat. Most people are conscious that they should eat healthy foods—high in protein, low in fat, containing the recommended daily allowance of vitamins and minerals, etc., etc. Those who care enough will carefully read the packages of foods in the local supermarket before purchasing them, but those who don't do so, remain unaware of the possibility that those foods have been somehow processed or modified from their original form. While the notion of foods containing preservatives and additives to enhance flavor and color is now a familiar concept to most Americans, few people realize that the food they purchase may have been modified on the genetic level. But just as the genetic engineering of animals has received more notoriety in recent years, so too have the methods by which farmers have engineered larger, redder tomatoes and sweeter ears of corn. For centuries, farmers have been altering the composition of plants in an attempt to grow sturdier, larger, and supposedly healthier agricultural products through cross-breeding and grafting. Nevertheless, as the articles in this section demonstrate, opposition to transgenic crops—crops grown from seeds that have been modified genetically in the laboratory—has steadily increased in recent years, with resistance coming from those who fear that, should they consume genetically modified foods, they will truly become what they eat (in their view, impure, corrupted organisms). Section VI focuses on the methods used to produce transgenic crops and the complex nature of the worldwide debate surrounding this growing technology.

The first article in this section is "The Global Food Fight," written by Robert Paarlberg for the journal *Foreign Affairs*. A comprehensive look at the causes and effects of the controversy over genetically modified (GM) crops, Paarlberg's essay explores the motives of its primary combatants, "cautious, consumer-driven Europe" and "aggressive American industry." After defining transgenesis, the process by which GM crops are bred, Paarlberg probes the ordeals of American corporations, including Monsanto, McDonald's, and Coca-Cola, when they encounter opposition from anti-GM activists in Europe and Asia. Caught in the middle are the developing nations for whom GM crops, engineered with resistance to various pests and diseases, are life-savers, and many Asian countries, which could benefit nutritionally from rice modified to contain essential vitamins otherwise lacking in their diet. Paarlberg concludes by calling on the policymakers in these countries to find a way to break their reliance on imported GM products and invest in their own technology.

145

Writer James Freeman next offers his own view on the debate in his article for *USA Today* entitled "You're Eating Genetically Modified Food." In his opinion, not only are GM foods inescapable, but we should not feel the need to escape them in the first place. By pointing out the pervasiveness of GM organisms, from our favorite breakfast cereal to the corn in our traditional Thanksgiving dinner, Freeman asserts that we have nothing to fear from consuming them. Meanwhile, in a point that echoes one implied by Paarlberg, Freeman explains how developing countries that grow GM crops have everything to gain from what he considers an essential tool of biotechnology.

The article which follows details a new procedure by which the genes in various organisms may be modified to benefit consumers while satisfying anti-GM activists. In an intriguing news item from the *New York Times* entitled "New Type of Gene Engineering Is Aimed at Sidestepping Critics," Barnaby J. Feder describes a form of technology by which organisms may be genetically engineered to mutate so that genes for disease tolerance already present in its DNA may be activated. The importance of producing rice engineered with essential vitamins is once again mentioned as a major benefit of this procedure, something that one opponent of GM foods calls "a noble effort."

The Global Food Fight[1]

By Robert Paarlberg
Foreign Affairs, May/June 2000

Food for Thought

Powerful new technologies often provoke strong resistance. When the internal combustion engine gave us automobiles, advocates of horse-drawn buggies scorned the fad. When nuclear fission was first mastered, much sentiment turned against its use—even for peaceful purposes. Thus today's backlash against the commercial use of recombinant DNA technology for food production should not be surprising. Consumer and environmental groups, mostly in Europe, depict genetically modified (GM) food crops, produced mostly in the United States, as dangerous to human health and the environment. These critics want tight labeling for GM foods, limits on international trade in GM crops, and perhaps even a moratorium on any further commercial development of this new technology—all to prevent risks that are still mostly hypothetical.

The international debate over GM crops pits a cautious, consumer-driven Europe against aggressive American industry. Yet the real stakeholders in this debate are poor farmers and poorly fed consumers in Asia, Africa, and Latin America. These are the regions most in need of new transgenic crop technologies, given their difficult farming conditions and rapidly growing populations. Yet poor farmers in tropical countries are neither participating in nor profiting from the GM crop revolution.

Gene Genie

The genetic modification of plants and animals through domestication and controlled breeding has gone on with little debate for roughly 10,000 years. But since 1973, genetic modification has also been possible through the transfer of isolated genes into the DNA of another organism. This type of genetic engineering—also known as genetic transformation, transgenesis, or simply GM—is a more powerful and more precise method of modifying life. Genes carrying specific traits can be transferred using a "gene gun" between species that would not normally be able to exchange

---

genetic material. A trait for cold resistance, for example, can be transferred from a fish to a plant.

As powerful as GM technology is, the large corporate investments needed to develop commercial applications for transgenic crops did not begin until 1980, when the U.S. Supreme Court extended patent protection to new types of plants and plant parts, including seeds, tissue cultures, and genes. Only after the Court guaranteed the protection of intellectual property rights did private corporations make the substantial investments necessary to develop commercially attractive transgenic crops.

The first GM crops that emerged were designed to solve important farm problems: pest control, weed control, and soil protection. The Monsanto Company, for example, developed soybeans with a built-in immunity to glyphosate, the active ingredient in the Monsanto herbicide Roundup. Having planted these GM soybeans, farmers could control weeds with a single spray of glyphosate, which had previously been lethal to the soybean plant. This reduced the need to employ more toxic and long-lasting weed killers or soil-damaging tillage. Several companies also developed GM varieties of cotton and corn engineered to contain a naturally occurring toxin— *Bacillus thuringiensis* (also known as Bt)—that minimizes insect damage to plants while dramatically reducing the need for chemical sprays.

> *The first GM crops that emerged were designed to solve important farm problems: pest control, weed control, and soil protection.*

These new GM field crops were finally released for large-scale commercial use by U.S. farmers in 1996. This followed years of laboratory testing and controlled field trials to screen for risks to other crops and animals, to the larger environment, and to human health. Once the Environmental Protection Agency, the Food and Drug Administration (FDA), and the U.S. Department of Agriculture approved the new GM seeds, American farmers gave them a try and instantly liked the results. By 1999, roughly half the U.S. soybean crop and one-third of the corn crop were genetically modified. While the seed companies made money, American farmers were the biggest winners, capturing roughly half of the total economic benefit from the new technology. (Patent-holders and seed companies gained only about a third of the added profits, while consumers got less than that.)

Enthusiasm for GM crops among American farmers is not hard to understand, given the decreased need for chemical sprays and tillage. Most U.S. farmers growing "Roundup Ready" soybeans need to spray only once, cutting chemical costs by 10-40 percent. Transgenic cotton often requires no spraying at all (compared to the 4-6

sprayings previously needed), reducing production costs by $60-
$120 per acre.

Surprisingly, however, the GM seed boom has only been effectively realized in three countries. In 1999, 72 percent of all land
planted with transgenics worldwide was in the United States,
while Argentina had 17 percent and Canada 10 percent. The nine
other countries that were (openly, at least) growing some transgenic crops—China, Australia South Africa, Mexico, Spain,
France, Portugal, Romania, and Ukraine—split the remaining one
percent.

The weak participation of tropical countries can be partly
explained by the industry's initial focus on temperate-zone crops
such as soybeans and corn. But how can we explain the lack of
enthusiasm among farmers in western Europe? There should have
been nothing to prevent these farmers from making the switch to
GM seeds. American companies have tried to market transgenic
seeds in Europe, and some attractive GM crops have also been
developed and patented by European-based companies. Yet within
the European Union the new technology has not taken hold. As of
1999, only a few farms in Spain, France, and Portugal were planting transgenic crops.

Allergic Reaction

European farmers have stayed away from transgenic crops
largely because European consumers have become frightened of
eating them. Consumers in Europe are now leading a backlash
against GM crops—even though no safety risks linked to any GM
crops on the market have ever been documented in Europe or any-
where else. After conducting its own 18-month study of this question, the U.K.-based Nuffield Council on Bioethics published the
following conclusion in May 1999:

> We have not been able to find any evidence of harm. We are
> satisfied that all products currently on the market have been
> rigorously screened by the regulatory authorities, that they
> continue to be monitored, and that no evidence of harm has
> been detected. We have concluded that all the GM food so far
> on the market in this country is safe for consumption.

Yet such expert reassurances are discounted by European consumers, distrustful since the 1996 "mad cow disease" scare. That
crisis undermined consumer trust in expert opinion after U.K.
public health officials gave consumers what proved to be a false
assurance that there was no danger in eating beef from diseased
animals. Although mad cow disease had nothing to do with the
genetic modification of food, it generated new consumer anxieties

about food safety at precisely the moment in 1996 when U.S.-grown GM soybeans were first being cleared for import into the EU.

Exploiting such anxieties, a number of third parties, including nongovernmental organizations (NGOS), quickly stepped into the fray. Greenpeace and other European activist groups that had previously struggled against nuclear power and the use of various man-made chemicals (especially chlorine, which Greenpeace had tried to label "the Devil's chemical") inflamed consumer phobias of GM foods. In Britain, Prince Charles (a self-described organic farmer) and Paul McCartney joined the chorus. In France—where food is never just food—a broad coalition of farmers, labor unions, environmentalists, and communists launched attacks against not only GM food but also McDonald's, imported beef grown with (non-GM) hormones, Coca-Cola, and various other threats to what they called French "culinary sovereignty." In Germany, GM opponents drew dark parallels between the genetic manipulation of food and their country's earlier lapse into human eugenics.

Although mad cow disease had nothing to do with the genetic modification of food, it generated new consumer anxieties about food safety....

These well-publicized campaigns forced significant corporate and government concessions in Europe. In April 1998, without scientific evidence of any harm from GM foods, Brussels stopped approving new GM crops for use in or import into the EU. This has meant a de facto ban on all corn imports from the United States (worth roughly $200 million annually), since bulk shipments might contain some GM varieties not yet approved. The EU also enacted a GM food labeling provision in 1998, requiring its 15 member states to begin marking all packaged foods that contain GM corn and soy. The United Kingdom went even further, requiring that restaurants, caterers, and bakers either list all GM ingredients or face fines of up to $8,400. To avoid consumer boycotts and lawsuits brought by activist groups, a growing number of food companies, retail stores, and fast-food chains (including both Burger King and McDonald's) in Europe pledged in 1999 not to use GM ingredients—at least where it could be avoided.

This backlash began to spread in 1999 to food-importing nations outside of Europe. Japan, South Korea, Australia, and New Zealand made plans to begin mandatory labeling for some transgenic foods, including heavily imported products such as GM soybeans and GM corn if intended for human consumption (as opposed to animal feed). Japan and South Korea together represent an $11.3 billion annual market for U.S. agriculture, and U.S. officials have worried that protectionist farm interests lie behind these labeling moves. But consumer anxiety is once again the more powerful factor at play. Responding to such fears, Japan's Kirin Brewery Company recently announced that starting in 2001 it would use only non-GM cornstarch for its beer; Kirin's competitor, Sapporo Breweries, made a similar announcement the next day.

Over Here

Europe's consumer-led backlash against GM crops put U.S. officials in an awkward spot. Usually the United States urges Europe and Japan to be more market-oriented in their food and agricultural policies; now, consumer-led market forces obliged the United States to adjust. U.S. officials have opposed the mandatory labeling of GM products. But the U.S. farm sector is so heavily export-oriented (U.S. farmers export more than 25 percent of the corn, soybean, and cotton they produce, and more than 50 percent of wheat and rice) that foreign pressure is prompting an informal movement in the other direction. The Archer Daniels Midland Company, a prominent U.S.-based soy-processing and export firm, announced in 1999 that it would henceforth ask U.S. farmers to deliver their GM and non-GM soybeans in separate batches so ADM could offer "GM free" products to consumers in Europe and Japan. Two large U.S.-based baby-food companies, Gerber and HJ. Heinz, announced in 1999 that they would soon switch to non-GM ingredients—not because of any new evidence that transgenic ingredients were unsafe, but out of fear of a Greenpeace-led boycott. Frito-Lay, the nation's major snack-food provider, followed suit, announcing that it would no longer use GM corn. In November, several members of Congress introduced a "Genetically Engineered Food Right to Know" bill that would require labels on any food containing at least 0.10 percent GM ingredients. The Grocery Manufacturers of America opposed this measure but supported stronger consultation requirements between food companies and the FDA, hoping to boost consumer confidence.

Credible labeling of all food produced from GM commodities would be an expensive proposition for U.S. farms, agribusinesses, and consumers. It would require complete physical segregation of GM and non-GM food along every step of production, from the farm gate to the grocery shelf. U.S. officials estimate that this could increase costs by 10-30 percent.

In the meantime, the European and Asian backlash against U.S.-grown GM crops could generate sharp conflicts in several international settings, including the World Trade Organization (WTO) and the Convention on Biological Diversity (CBD). Within the WTO, the Sanitary and Phytosanitary (SPS) Agreement permits nations to restrict imports in the name of health or environmental protection. But an unresolved question is whether governments can restrict imports under conditions of scientific uncertainty, on a precautionary basis. The SPS agreement allows import restrictions only on a provisional basis while governments seek additional information.

The EU is trying to weaken this WTO requirement. In January 2000, it managed to insert language supporting its precautionary principle into the text of the new Protocol on Biosafety in the CBD. Hammered out by environmental rather than trade ministers, this protocol was drafted specifically to govern international trade in transgenic organisms, and it now states in several places that a "lack of scientific certainty due to insufficient relevant scientific information and knowledge" should not prevent states from taking precautionary import actions. The protocol then goes on to oblige exporters of living modified organisms meant for environmental release (such as plants or seeds) to provide prior notification of relevant biosafety information and to solicit an informed consent agreement from importers.

> *Credible labeling of all food produced from GM commodities would be an expensive proposition for U.S. farms, agribusinesses, and consumers.*

The United States fought to include language in the protocol that would place it under the authority of WTO rules, but was blocked from doing so by the EU and most developing countries. State Department officials reluctantly accepted the final terms of the protocol, partly with the hope that it might calm consumer and importer fears if the United States and the EU were seen to agree on the issue. By accepting the protocol, the United States also avoided further isolation within the CBD (to which Washington is not yet a formal party, since the Senate has failed to ratify it). But this acquiescence may have weakened America's hand on future GM trade issues within the WTO.

Such conflicts between the United States and Europe over GM crops may continue to escalate in the months and years ahead. Yet the most important stakeholders in the fight over GM foods have not been heard. It is among poor farmers and poor consumers in developing countries that the potential gains from this new technology are most significant. In the tropics, many consumers are not yet well-fed and most farmers are not yet wealthy. Larger investments in the genetic modification of some crops could open a new avenue of escape from poverty and malnutrition for hundreds of millions of citizens in Asia, Africa, and Latin America. Yet far too little is being done to make that happen.

Serious Stakes

If properly exploited, the GM crop revolution will have life changing—and even live-saving—implications in developing countries. Food-production requirements are increasing rapidly in the tropics due to population growth. Yet agriculture there is lagging, in part because of poor soil; extremes of moisture, heat, and drought; and a plenitude of pests and diseases that attack animals and crops. Poor farmers in tropical Asia and Africa currently lose much of their crop production every year (often more than 30 percent) to insects and plant disease.

Here is where modern transgenic technology carries special promise for the tropics: it can engineer plants and animals with highly specific pest and disease resistances. For example, poor farmers in Kenya today lose 15-45 percent of their maize to stem borers and other insects. If they could plant maize seeds engineered to contain Bt, a pest-killing toxin, they could reduce their losses without reliance on chemical sprays. Similarly, transgenic virus-resistant potatoes could help small-scale farmers in Mexico who currently suffer substantial crop damage. And a World Bank panel has estimated that transgenic technologies could increase rice production in Asia by 10-25 percent within the next decade. Without such gains, increasing demand from a growing population could push the price of rice beyond the reach of the poor.

If properly exploited, the GM crop revolution will have life changing-and even live-saving-implications in developing countries.

Genetic technology could also improve nutrition. If the 250 million malnourished Asians who currently subsist on rice were able to grow and consume rice genetically modified to contain Vitamin A and iron, cases of Vitamin A deficiency (which currently kills 2 million a year and blinds hundreds of thousands of children) would fall, as would the incidence of anemia (one of the main killers of women of childbearing age).

The U.N.'s Food and Agriculture Organization has recently estimated that one out of every five citizens of the developing world— 828 million people in all—still suffers from chronic undernourishment. One reason for this is lagging agricultural production in some poor regions despite the earlier innovations of the so-called green revolution. The disadvantaged (and mostly female) farmers of Africa were bypassed by the dramatic gains brought on by the conventional (non-GM) plant-breeding breakthroughs of the 1960s and 1970s. Between 1970 and 1983, new high-yielding rice varieties spread to about 50 percent of Asia's vast rice lands but to only about 15 percent in sub-Saharan Africa. Similarly, improved wheat varieties spread to more than 90 percent of Asia and Latin

America but to only 59 percent of sub-Saharan Africa. This helps explain why agricultural production has increased ahead of population growth in both East and South Asia while falling behind population growth in sub-Saharan Africa—leaving an estimated 39 percent of Africans undernourished.

African farmers fell behind because they had greater difficulty than Asians in getting access to the full package of green revolution technology. Earlier cross-bred crops still required farmers to buy supplementary products, such as chemical sprays. But with new transgenic crops, all the potential for enhanced productivity exists in the seed itself. Pests and diseases are managed not with chemicals but through genetic engineering.

Critics of the GM revolution fear that the environment might be hurt if engineered crops are released into rural tropical settings where wild relatives of food plants can often be found. If an engineered herbicide-resistance trait breeds into a weedy wild relative, the result might be a hard-to-manage "superweed." Or widespread planting of Bt crops might trigger an evolving population of "superbugs" resistant to the toxin. Legitimate biosafety concerns such as these have so far been addressed in rich countries on a case-by-case basis, through field testing under closely monitored conditions; the means for such testing and monitoring are still largely missing in the developing world. Even so, the hypothetical threat to biosafety posed by GM crops remains demonstrably smaller than the actual threat posed by invasions of exotic but non-GM plant and animal species. By some estimates, exotic species movements (having nothing to do with genetic engineering) currently generate tens of billions of dollars in losses to agriculture annually in the developing world. If these countries are truly concerned with biosafety, GM crops should hardly be their first focus.

Transgenic products not only reduce chemical sprays, they can also aid in land conservation and species protection. For small farmers in the tropics, if GM crops or animal vaccines make farm and grazing lands more productive, there will be less need to plow up or graze more fragile lands in the future. In sub-Saharan Africa, roughly 5 million hectares of forest are lost every year, primarily to new clearance for low-yield agriculture. The real threat to biodiversity in poor countries today comes from such cutting of natural habitats. Thus the ultimate environmental pay off from transgenic crop technologies could include fewer watersheds destroyed, fewer hillsides plowed, fewer trees cut, and more species saved.

Pound Foolish

Although the GM crop revolution could greatly benefit poor farmers in poor countries, this potential is not being realized. As noted above, their relatively prosperous colleagues in North America and Argentina grow 99 percent of all GM crops. Why have poor farmers in developing countries not participated in the boom?

Transgenic products not only reduce chemical sprays, they can also aid in land conservation and species protection.

First, consider the market-driven motives of the private GM seed companies that have been making the largest investments in this new technology. These multinationals have been criticized for their alleged efforts to make poor farmers in the developing world dependent on GM seeds. In fact, the GM seeds these companies are bringing to market have mostly been designed for sale to farmers in rich (mostly temperate-zone) countries. The danger is not that poor farmers in the tropics will become dependent on these companies; the danger is that corporate investments will mostly ignore the tropics because farmers there do not have the purchasing power to buy expensive GM seeds.

Some GM crop technologies originally developed for the temperate zone (Bt maize and cotton, for example) might readily be adapted for use in the tropics by transferring the desirable GM traits into locally grown crops through conventional plant breeding. Private companies, however, have little incentive to invest in such local adaptations where farmers are poor. Worse, they may seek to block local adaptations if poor counties are not willing to protect corporate intellectual property rights (IPRS). Seed companies had once hoped to solve property problems by engineering a natural sterility (called gene-use restriction technology, or GURT) into the seeds of GM plants. But such thoughts were set aside in 1999 when Monsanto agreed, under intense pressure from critics, not to commercialize its "terminator" GURT technology.

Protection of intellectual property is less of a problem in rich countries such as the United States. If anything, the U.S. Patent and Trademark Office has given corporations more protection than is good for them. Companies can now patent not just the inventive use of plant traits and genes, but also some of the smallest fragments of genetic material. Since the commercialization of a single transgenic insect-tolerant plant can now require the combination of many separately patented subtechnologies, problems with legal gridlock arise.

In most developing countries, however, IPR protection for GM crops tends to be too weak rather than too strong. A WTO agreement on trade-related aspects of intellectual property rights (TRIPS), reached during the Uruguay Round of negotiations, requires that all WTO members—including even the poorest countries after 2006–provide IPR protection for plant varieties. Yet many developing countries will try to satisfy TRIPS without giving up the traditional privileges of farmers to replicate and replant protected seeds on their farms.

This being the case, corporations will remain wary. As long as both purchasing power and IPR protection remain missing, private firms will probably not invest in the innovations most needed by poor farmers in tropical countries. For these farmers, the marketplace by itself is unlikely to produce much GM magic. Market forces have not prompted international drug companies to do adequate research in tropical diseases such as malaria. Similarly, market forces alone will not trigger the GM crop investments most needed by poor farmers in Asia and Africa.

Companies can now patent not just the inventive use of plant traits and genes, but also some of the smallest fragments of genetic material.

A historical comparison drives the point home. Hundreds of millions of poor farmers in the developing world (at least those on good land) benefited from the earlier green revolution because in that case private multinationals were not in the lead. Instead, the leaders were governments, international financial institutions, and private philanthropies (especially the Ford and Rockefeller foundations). Market-oriented corporations did not build the laboratories or support the plant-breeding efforts in Mexico and in the Philippines that led to new, high-yielding varieties of wheat and rice in the 1950s and 1960s. These strains were developed and later adapted for local use by plant breeders working within the public sector, paid for in large part by Cold War-era foreign aid. The adapted local varieties were then replicated by national seed companies and given away to farmers. Intellectual property rights were not an issue, since government agencies wanted the seeds to spread as fast as possible. During this original green revolution, the public sector often went so far as to extend subsidies to farmers for cheap irrigation and fertilizers along with the seeds themselves.

Today's public-sector institutions are showing much less leadership in promoting the gene revolution. Reasons for this include a mistaken impression that all regions shared in the green revolu-

Today's public-sector institutions are showing much less leadership in promoting the gene revolution. Reasons for this include a mistaken impression that all regions shared in the green revolution's success; the much larger and riskier investments in science that are needed to develop and commercialize new GM crop varieties; the dramatic shrinkage in budget leeway in most developing countries since the 1980s debt crisis; the model of market-led development pushed onto borrowing countries by the World Bank and the International Monetary Fund after that crisis; the disrepute of public sector-led development following the collapse of the Soviet Union; and finally, the diminished rationale for generous foreign aid to poor countries following the end of the Cold War.

Unfortunately, public development institutions also shy away from investment in GM technology out of fear: fear of media criticism, of litigation, or of physical attack by anti-GM activists. These are not imagined risks. The headquarters of the U.S. Agency for International Development's principal developing-country biotechnology support project, located at Michigan State University, was set on fire just before midnight on December 31, 1999, by an underground group calling itself the Earth Liberation Front.

More than just GM research is being left undone. Public-sector support for agricultural development has collapsed across the board. Annual foreign aid to agriculture in poor countries fell by 57 percent between 1988 and 1996 (from $9.24 billion down to just $4.0 billion, measured in constant 1990 dollars), and annual World Bank lending for agriculture and rural development fell by 47 percent between 1986 and 1998 (from $6 billion to just $3.2 billion, measured in constant 1996 dollars). As donors have pulled back, governments in the developing world have not filled the gap. Poor countries remain notoriously unmindful of the need to invest in agriculture, despite the documented high payoffs. These governments are distracted by demands from more powerful urban constituencies, often led by the army, state-owned industries, or the state bureaucracy. On average, developing countries devote only 7.5 percent of total government spending to agriculture, and little of this goes for research. Sub-Saharan Africa has only 42 agricultural researchers per million economically active persons in agriculture, compared with an average of 2,458 researchers per million in developed countries.

Even taking these private-sector limitations and public-sector lapses into account, the near total exclusion of poor-country farmers from today's GM crop revolution remains surprising. Even where useful GM technologies are commercially available, officials in poor countries have been curiously slow to allow their use. One reason has been the export to the developing world of the highly cautious attitude of European consumers and environmental groups toward GM crops. European fears have been exported both through market channels and through activist campaigns launched or supported by European-based NGOS.

In Thailand, for example, where exports of agricultural products such as rice, shrimp, tapioca, and poultry provide 23 percent of total export earnings and where local scientists have already engineered some improved GM crop varieties under greenhouse conditions, the actual planting of GM seeds is now blocked by the government. Warnings from customers in Europe and Australia that Thai exports might be shunned if they include any GM ingredients prompted Bangkok to announce, in mid-1999, that henceforth GM seeds would not be brought into the country until proven safe for human consumption. Some GM soybean and cotton seeds (grown safely and profitably by farmers in the United States since 1996) are rumored to be reaching Thai farmers through black-market channels, but the Thai government—which until recently had supported GM crops—now views such imports as criminal.

Poor countries remain notoriously unmindful of the need to invest in agriculture, despite the documented high payoffs.

In Brazil, farmers who had hoped to plant herbicide-resistant soybeans in 1999 were blocked at the last moment when a federal judge granted an injunction filed by Greenpeace and a Brazilian consumer institute on grounds of a possible threat to the Brazilian environment. Higher courts are now reviewing the case, but a ban on planting remains in place. Farmers eager to get GM soybean seeds have been smuggling them in from Argentina, but the state government of Rio Grande do Sul, partly in hopes of being able to offer GM-free products to customers in Europe and Japan, has threatened to burn their fields and jail any farmers found to be growing GM soybeans. Greenpeace has thrown its weight behind efforts to keep Rio Grande do Sul a "GM-free zone."

In India, devastating bollworm infestations in cotton plants have brought despair—and reportedly hundreds of suicides—to poor cotton farmers. Insects have developed resistance to the heavy volume of pesticides sprayed on Indian fields. (Cotton accounts for 50 percent of all pesticide sprayed in India, even though the crop takes up only 5 percent of total farmland.) In recent Indian field tests, a GM cotton variety genetically modified to control bollworm increased crop yields by 40 percent while permitting seven fewer sprayings. But commercial release has been delayed because NGOS have filed a public-interest lawsuit against the government agency that authorized the trials, and activists have destroyed some of the test fields. Many of the same activist groups that oppose GM seeds in India today also opposed the introduction of improved non-GM seeds during the earlier green revolution.

Tragically, the leading players in this global GM food fight—U. S.-based industry advocates on the one hand and European consumers and environmentalists on the other—simply do not reliably represent the interests of farmers or consumers in poor countries. With government leadership and investment missing, the public interest has been poorly served. When national governments, foreign donors, and international institutions pull back from making investments of their own in shaping a potentially valuable new technology, the subsequent public debate naturally deteriorates into a grudge match between aggressive corporations and their most confrontational NGO adversaries. This confrontation then frightens the public sector, deepening the paralysis.

Breaking that paralysis will require courageous leadership, especially from policymakers in developing countries. These leaders need to carve out a greater measure of independence from the GM food debate in Europe and the United States. Much larger public-sector investments of their own in basic and applied agricultural research will be necessary to achieve this autonomy. New investments in locally generated technology represent not just a path to sustainable food security for the rural poor in these countries; in today's knowledge-driven world, such investments are increasingly the key to independence itself.

You're Eating Genetically Modified Food[2]

BY JAMES FREEMAN
USA TODAY, FEBRUARY 9, 2000

There's no escape. You are consuming mass quantities of genetically modified food. The milk on your Cheerios this morning came from a genetically modified cow, and the Cheerios themselves featured genetically modified whole grain goodness. At lunch you'll enjoy french fries from genetically modified potatoes and perhaps a bucket of genetically modified fried chicken. If you don't have any meetings this afternoon, maybe you'll wash it all down with the finest genetically modified hops, grains and barley, brewed to perfection—or at least to completion if you're drinking Schaefer.

Everything you eat is the result of genetic modification. When a rancher in Wyoming selected his stud bull to mate with a certain cow to produce the calf that ultimately produced the milk on your breakfast table, he was manipulating genes. Sounds delicious, doesn't it? Sorry, but you get the point.

Long before you were ever born, farmers were splicing genes, manipulating seeds to create more robust plants. Genetic modification used to be called "breeding," and people have been doing it for centuries. Thomas Jefferson did it at Monticello, as he experimented in his gardens with literally hundreds of varieties of fruits and vegetables. Hmm, Thomas Jefferson and genes . . . his column is going to disappoint a lot of people doing Web searches.

Anyway, to return to the topic at hand, breeding isn't a scary word, so people who oppose technology call it "genetic modification." They want to cast biotechnology, which is just a more precise and effective breeding tool, as some kind of threat to our lives, instead of the blessing that it is.

Have you ever seen corn in its natural state without genetic modification? It's disgusting. We're talking about that nasty, gnarled, multi-colored garbage used as ornamentation in Thanksgiving displays. The fear mongers should eat that the next time they want to criticize technology.

In fact, the fear mongers are waging a very successful campaign against biotechnology, especially in Europe where they've lobbied to limit the availability of "genetically modified" foods. Even in the

---

2. James Freeman writes a weekly opinion column for *USAToday.com* and a weekly TechnoPolitics column for *Forbes.com*.

United States, where we generally embrace technology and its possibilities, the fear is spreading. Not because of some horrible event related to the food supply, but because of more aggressive spinning of the media. In fact, you've been enjoying foods enhanced by biotechnology for most of the last decade. And the news is all good—lower prices and more abundant food.

As for the future, the potential to eliminate human suffering is enormous. Right now, according to the World Health Organization, more than a million kids die every year because they lack Vitamin A in their diets. Millions more become blind. WHO estimates that more than a billion people suffer from anemia, caused by iron deficiency. What if we could develop rice or corn plants with all of the essential vitamins for children?

Personally, I'd rather have an entire day's nutrition bio-engineered into a Twinkie or a pan pizza, but I recognize the benefits of more-nutritious crops. Reasonable people can disagree on the best applications for this technology.

Still, the critics want to talk about the dangers of genetically modified crops. The Environmental Protection Agency wants to regulate the use of certain bio-engineered corn seeds because they include a resistance to pests. Specifically, the seeds are bred to include a toxin called BT that kills little creatures called corn borers, so farmers don't need to spray pesticides.

Turns out, according to the EPA, that the toxin in the corn can kill Monarch butterflies, too. The butterflies don't eat corn, but the EPA is afraid that the corn pollen will blow over and land on a milkweed and stick to it and then confused Monarch caterpillars will inadvertently eat the pollen.

. . . you've been enjoying foods enhanced by biotechnology for most of the last decade. And the news is all good— lower prices and more abundant food.

Not exactly the end of the world, but it sounds bad. Until you consider the alternatives. According to Professor Nina Fedoroff, Willaman Professor of Life Sciences at Penn State, "A wide-spectrum pesticide sprayed from a plane is going to kill a lot more insects than will be killed by an in-plant toxin."

Of course, the anti-tech crowd will say that they don't like pesticides either. They promote organic farming—meaning we use more land to produce our food, and we clear more wilderness. We also pay more for food, since we're not using the efficiencies that come from technology. Maybe that's not a problem for you or me, but it's bad news for those millions of malnourished kids around the world.

Says Fedoroff, "I think that most inhabitants of contemporary urban societies don't have a clue about how tough it is to grow enough food for the human population in competition with bacteria, fungi, insects and animals and in the face of droughts, floods, and other climatic variations."

That may be true, but I do think that most Americans understand the positive impact of technology. And that's why they'll ultimately reject the scare campaign against biotechnology.

New Type of Gene Engineering Is Aimed at Sidestepping Critics[3]

By Barnaby J. Feder
NEW YORK TIMES, February 29, 2000

Researchers estimate that rice and corn began evolving their separate ways from a common grassy ancestor at least 60 million years ago. But how far did they travel?

Intriguingly, researchers have found that the genes in corn that tell the plant what proteins to make to produce its shape, its cob, its roots and its reproductive system—in short, everything that makes it corn—have largely identical counterparts in rice, arrayed in pretty much the same order.

That suggests that the differences between the plants might come in large part not from the genes themselves, but from the point at which they are switched on and off, how strongly and in which part of the plant they are active. And now some wonder if this insight can be used for a new kind of genetic engineering.

No one is more enthusiastic about such inquiries than Dr. Richard A. Jefferson, the 43-year-old founder of the Center for the Application of Molecular Biology to International Agriculture in Canberra, Australia, who sometimes declares, "Rice is corn." Although he knows that is an exaggeration, it serves as a catchy introduction to how his institute is trying to reshape the tangled global debate over the role of genetic engineering in agricultural biotechnology.

Dr. Jefferson argues that the high degree of genetic overlap between the plants—indeed, among all living things—suggests that much of the gene swapping among species that has stirred up so much opposition to genetic engineering may be unnecessary. Perhaps, instead of moving a valuable trait like resistance to cold from a fish to a plant, genetic engineers could achieve the same result by goading the plant into a mutation that activates genes for cold tolerance already present in its DNA.

Dr. Jefferson, a native of Berkeley, Calif., is not opposed to moving genes among different forms of life. In fact, he first made his mark in biotechnology at the University of Colorado in 1985 by inventing a way to track the location and activity of genes as researchers moved them from one species to another. In 1987, in one of the first field tests of a genetically altered food crop, he raised genetically altered potatoes at the Plant Breeding Institute in Cambridge, England.

Today, though, Dr. Jefferson's nonprofit research center, known as Cambia, is pouring resources into a project to jumble and rejumble the on-off patterns of rice genes, hoping to unleash traits buried in rice that evolution might not get around to exhibiting for millions of years, if ever.

Promising mutants—say rice that produces vitamin A in the grain—could be developed into viable crops in developing countries within four or five years by crossing them with existing crops, according to Dr. Andrzej Kilian, who was hired to run the research. And Cambia is working with partners to apply the same concepts to cassava, cowpea and other plants vital to food supplies in developing countries.

Laboratory-driven mutation work is just one of many Cambia projects aimed at the needs of farmers in developing countries.

"Evolution uses random forces all the time," Dr. Kilian said. "We are trying to speed it up and make it manageable."

Laboratory-driven mutation work is just one of many Cambia projects aimed at the needs of farmers in developing countries. Dr. Jefferson hopes such efforts will create a way for small businesses and developing countries to exploit biotechnology while skirting the fortress of patents that Monsanto, DuPont and other multinational giants have assembled.

And that, in turn, could please some current critics of biotechnology.

"It's a noble effort," said Hope Shand, research director of the Rural Advancement Foundation International U.S.A., a group based in Pittsford, N.C., which has been a high-profile opponent of the biotechnology industry. "Our main concern has always been who benefits and who controls the technology."

Cambia's technology still crosses species lines in ways that could upset some critics of biotechnology. Cambia's process, which it calls transgenomics, does not move genes intended to introduce a novel trait into rice, but its method for changing rice's regulation of its own genes does require on-off switches to be imported from bacteria, yeast or other plants.

"People may look at this differently than some of the transspecies work but some of the risks they worry about would still be there," said Dr. Margaret Mellon, who tracks biotechnology issues at the Union of Concerned Scientists in Washington.

Dr. Mellon said that Cambia's work might provide valuable research on plant genetics, but that those concerned with world hunger ought to see it as a poor alternative to breeding programs crossing rice with wild relatives, like that being run in China by Chinese researchers and Dr. Susan McCouch, a Cornell scientist.

Dr. McCouch has reported increasing yields on commercial rice strains by 10 percent to 20 percent by crossing them with wild species. In one case, the offspring surprisingly proved to be resistant to a virus plaguing Latin American rice growers, even though neither parent was.

Such work is often cited by opponents of biotechnology as proof that genetic engineering is unnecessary and distracting, but Dr. McCouch herself doubts that traditional breeding alone will meet the developing world's food needs. Dr. McCouch and others say the transgenomics program is a logical, if bold, use of the growing mountain of research on how mutations occur naturally.

It has been known for decades that organisms occasionally cut loose pieces of their own regulatory DNA so that they can jump, apparently randomly, to other locations and possibly activate genes with helpful traits. Production of the enzymes, or transposases, that free the mobile DNA, or transposons, increases when organisms are under almost any kind of stress.

"One of the genome's last ditch responses under stress is to reshuffle the deck," Dr. McCouch said.

Cambia is trying to mimic the natural process by inserting a packet of DNA from yeast, bacteria and corn DNA that can only be activated when a rice gene is nearby and active. When the rice gene is switched on, it simultaneously turns on a cascade of activity in the inserted DNA packet: one result is that a "reporter" gene—Cambia's comes from bacteria—begins producing a protein that has no effect on the plant but can be used by researchers to figure out where the packet has landed in the rice genome and how powerfully the nearby rice gene is working.

> *Dr. [Susan] McCouch and others say the transgenomics program is a logical, if bold, use of the growing mountain of research on how mutations occur naturally.*

The activity also turns on DNA fragments that Cambia takes from yeast that amplify gene activity. That can boost the activity of the adjacent rice gene, which may produce noticeable changes in the plant. But Cambia's real mutation thrust is based on including in the DNA packet a DNA package that can become mobile when the plant is crossed with another plant containing an enzyme that frees it.

If the mobile segment jumps from, for example, a location next to a rice gene active during seed development to the neighborhood of a gene involved in root growth, it will be turned on and stimulate the root gene when the seed gene becomes active.

That process mimics what a rice transposon might do naturally if the plant were under stress. Researchers must still screen thousands of plants, just as natural breeders do, looking for mutations that are beneficial to farmers.

Cambia is now building a library of plants with the DNA packages inserted near various rice genes. It has about 3,000 rice plants and expects to reach a goal of 10,000 by the end of the year.

Theoretically, there is plenty of room to create new functions for existing genes. "Plant genes are generally redundant," said Dr. Susan Wessler, an expert in corn and rice genetics at the University of Georgia. "There may be eight copies of a gene." Thus, one copy of the gene could be reregulated to produce a new trait by a transposon without necessarily compromising the plant.

Multinational companies have been notably silent about whether they are doing similar research. Dr. Jefferson said some were negotiating nonexclusive licensing agreements for transgenomics technology, parts of which could be used to make existing genetic engineering more precise.

For instance, corn might be engineered so that its bacterially derived pesticide gene is turned on only in the parts of the plant that corn borers eat. Now, engineered corn expresses the pesticide all over the plant, including in pollen that can be deadly to monarch butterflies and other beneficial insects.

But plenty of admirers working in research are hoping Dr. Jefferson's aim of using transgenomics to perform an end run around the multinationals' current approach to biotechnology pays off.

"Richard's focus is on those people for whom agriculture is most vital," said Dr. Jeffrey Bennetzen, a Purdue University plant genetics researcher. "Our biggest problem is   overproduction, but the developing world really needs crop improvement."

Bibliography

Books

Aldridge, Susan. *The Thread of Life: The Story of Genes and Genetic Engineering*. New York: Cambridge UP, 1996.

Anderson, Luke. *Genetic Engineering, Food, and Our Environment*. White River Junction, VT: Chelsea Green Pub. Co., 2000.

Bains, William. *Biotechnology from A to Z*. New York: Oxford UP, 1998.

Bosk, Charles L. *All God's Mistakes: Genetic Counseling in a Pediatric Hospital.*Chicago: U of Chicago P, 1992.

Bud, Robert. *The Uses of Life: A History of Biotechnology*. New York: Cambridge UP, 1993.

Cole-Turner, Ronald, ed. *Human Cloning: Religious Responses*. Louisville, KY: Westminster/John Knox P, 1998.

Dawkins, Kristin. *Gene Wars: The Politics of Biotechnology*. New York: Seven Stories P, 1997.

Du Charme, Wesley M. *Becoming Immortal: Nanotechnology, You, and the Demise of Death*. Evergreen, CO: Blue Creek Ventures, 1995.

Grace, Eric S. *Biotechnology Unzipped: Promises & Realities*. Washington, D.C.: Joseph Henry P, 1997.

Heinberg, Richard. *Cloning the Buddha: The Moral Impact of Biotechnology*. Wheaton, IL.: Quest Books, 1999.

Human Genome Program, U.S. Department of Energy.

——. *To Know Ourselves*. Washington, D.C., 1996.

Kaplan, Jonathan Michael. *The Limits and Lies of Human Genetic Research*. New York: Routledge, 2000.

Kimbrell, Andrew. *The Human Body Shop: The Cloning, Engineering, and Marketing of Life*. Washington, D.C.: Regnery Pub., 1997.

Kitcher, Philip. *The Lives to Come: The Genetic Revolution and Human Possibilities*. New York: Simon & Schuster, 1997.

Kolata, Gina. *Clone: The Road to Dolly and the Path Ahead*. New York: William Morrow & Co., 1998.

Lanza, R. P., David K. C. Cooper, and Robin Cook. *Xeno: The Promise of Transplanting Animal Organs into Humans*. New York: Oxford UP, 2000.

Lewontin, Richard. *The Tripple Helix: Gene, Organism, and Environment*. Cambridge, MA: Harvard UP, 2000.

Lyon, Jeff, and Peter Gorner. *Altered Fates: Gene Therapy and the Retooling of Human Life*. New York: Norton, 1995.

Marshall, Elizabeth L. *High-Tech Harvest: A Look at Genetically Engineered Foods*. New York: Franklin Watts, 1999.

McGee, Glenn, ed. *The Human Cloning Debate*. Berkeley, CA: Berkeley Hills Books, 2000.

McGee, Glenn. *The Perfect Baby: A Pragmatic Approach to Genetics*. Lanham: Rowman & Littlefield Publishers, 1997.

McKelvey, Maureen D. *Evolutionary Innovations: The Business of Biotechnology*. New York: Oxford UP, 2000.

Murphy, Timothy F., and Marc A. Lappé, eds. *Justice and the Human Genome Project*. Berkeley: U of California P, 1994.

Nottingham, Stephen. *Eat Your Genes: How Genetically Modified Food Is Entering Our Diet*. New York: Zed Books Ltd, 1998.

Oliver, Richard W. *The Coming Biotech Age: The Business of Bio-Materials*. New York: McGraw Hill, 2000.

O'Mahony, Patrick, ed. *Nature, Risk and Responsibility: Discourses of Biotechnology*. New York: Routledge, 1999.

Peters, Ted. *For the Love of Children: Genetic Technology and the Future of the Family*. Louisville, KY: Westminster/John Knox P, 1996.

——. *Playing God?: Genetic Determinism and Human Freedom*. New York: Routledge, 1996.

Regis, Ed. *Nano! The Emerging Science of Nanotechnology: Remaking the World—Molecule by Molecule*. Boston: Little, Brown, 1995.

Ridley, Matt. *Genome: The Autobiography of a Species in 23 Chapters*. New York: HarperCollins, 2000.

Rifkin, Jeremy. *The Biotech Century: Harnessing the Gene and Remaking the World*. New York: Jeremy P. Tarcher/Putnam, 1998.

Robbins-Roth, Cynthia. *From Alchemy to IPO: The Business of Biotechnology*. Cambridge, MA: Perseus P, 2000.

Rudolph, Frederick B., and Larry B. McIntire, eds. *Biotechnology: Science, Engineering, and Ethical Challenges for the Twenty-first Century*. Washington, D.C.: Joseph Henry P, 1996.

Wilmut, Ian, Keith Campbell, and Colin Tudge. *The Second Creation: Dolly and the Age of Biological Control*. New York : Farrar, Straus and Giroux, 1999.

Wright, William. *Born That Way: Genes, Behavior, Personality*. New York: Routledge, 1999.

Additional Periodical Articles with Abstracts

Those interested in reading more about issues in biotechnology may refer to the following list of articles. Readers who require a more comprehensive selection are advised to consult *Reader's Guide Abstracts* and other H. W. Wilson indexes.

DNA Vaccines as Cancer Treatment. Edward P. Cohen. *American Scientist* v. 87 pp328-35 July/Aug 1999.

Cohen examines scientists' belief that the power of the immune system can be marshaled with the use of vaccines to treat cancer. He explains that the human immune system is already known to attack cancer cells but is not strong enough to kill all cancer cells. According to Cohen, researchers are currently exploring the possibility that a vaccine derived from tumor DNA could be used to augment the immune response to such a degree that dangerous cancerous cells could be successfully eliminated.

Transgenic Foods: Promise or Peril? Anne Acosta. *Americas* v. 52 pp14-15 May/June 2000.

As Acosta explains, transgenic food is one of the most controversial areas of scientific study. Developments in the field of genetics over the past two decades have enabled scientists to locate specific genes that can be moved from one species to another and between viruses, bacteria, plants, and animals to create significant modifications in the host species. According to Acosta, genetically altered food, with its potential to increase the world's food supply, offers great promise to fulfill one of the world's basic needs. Nonetheless, genetically modified food raises concerns for many people with regard to safety, monopoly control over basic foodstuffs, and the implications of genetic manipulation for biodiversity and for the human race.

Gene Researchers, Hold Your Hype. Paul Raeburn. *Business Week* p88 April 3, 2000.

Raeburn writes of the competition to develop a new class of gene therapies and genetically engineered drugs that could revolutionize the treatment of heart disease. He reports that so-called "angiogenesis" drugs encourage the growth of new blood vessels in the heart, offering a detour around clogged arteries; early results suggest that they could eventually offer an important alternative to the 500,000 bypass operations and 400,000 angioplasties performed in the United States every year. Raeburn says, however, that at the

2000 conference of the American College of Cardiology in Anaheim, California, researchers distorted and inflated inconclusive test results. Raeburn concludes that, though the new treatments will probably be very successful, progress relies on the sober assessment of the latest results rather than on hype.

Score One for Gene Therapy. Ellen Licking. *Business Week* pp58-60 May 8 2000.

Licking reports that a team of French researchers has made a breakthrough in gene therapy. According to the April 28, 2000, issue of *Science*, researchers at the Necker hospital in Paris said they had successfully treated two infants with a fatal immune disease. The biologists harvested the bone marrow of two babies with severe combined immunodeficiency X1, isolated its stem cells, and bathed them in biological factors in order to encourage the cells to adopt a correct version of the diseased gene. Although experts say it is too early to declare victory, Licking says, the results of the treatment are very encouraging. According to Licking, R. Michael Blaese, head of human therapy at biotech firm ValiGen, says that if the technique is a success it could be used to treat a range of diseases, including AIDS and other anemias.

Could Better Reports by Researchers Have Prevented a Clinical-trial Death? Jeffrey Brainard. *Chronicle of Higher Education* v. 46 ppA45-A46 April 14, 2000.

Brainard probes the 1999 death of a young person in a gene therapy trial, which has raised several issues for scientists and government officials. Among them, he says, are the questions of when clinical researchers should report that medical research participants suffered unusual health problems and whether the institutional review boards and federal agencies that examine adverse events reports are equipped for that role.

The Frozen Zoo. *Discover* v. 20 p21 Oct 1999.

This article profiles the work of Duane Kraemer, a professor of veterinary physiology at Texas A&M University, who is leading an effort to deep-freeze genetic material from 2,000 species in danger of extinction as part of Project Noah's Ark. In the future, it is explained, extinct animals could be cloned back to life to rebuild wild populations or reduce inbreeding in captivity. In the article, Kraemer stresses both the difficulties and importance of obtaining cells from at least 200 pandas—only 1,000 of which survive—before they die out, while critics fear that cloning efforts lessen the motivation to protect threatened ecosystems, such as China's bamboo forests.

The Environmental Genome Project: Ethical, Legal, and Social Implications. Richard R. Sharp. *Environmental Health Perspectives* v. 108 pp279-81 April 2000.

Sharp reports on the National Institute of Environmental Health Sciences' support of a multiyear research initiative examining genetic influences on environmental response. According to Sharp, proponents of this new initiative, known as the Environmental Genome Project, hope that new information will improve our understanding of environmentally associated diseases and allow clinicians and public health officials to target disease-prevention strategies to those who are at increased risk. Despite these potential benefits, Sharp explains, the project presents several challenges, including the protection of individual research participants, ethical issues related to the application of research results, and how study findings could affect social priorities. Sharp states that clarifying these concerns, many of which have not received adequate attention from bioethicists, is essential to minimize potential research-related risks and define research needs.

Disability, Gene Therapy and Eugenics—A Challenge to John Harris. Solveig Magnus Reindal. *Journal of Medical Ethics* v. 26 pp89-94 April 2000.

Reindal challenges the view of disability presented by John Harris in his article, "Is gene therapy a form of eugenics?" He argues that Harris's definition of disability rests on a model wherein disability is regarded as a product of biological determinism or "personal tragedy" in the individual. Reindal explains that this view, often called "the medical model," has been criticized for dealing more often with the term impairment than "disability." The shortcomings of the individual model of disability are stated and it is argued that an adequate ethical discourse on gene therapy requires that perspectives from disability research be taken into consideration.

Should We Clone Human Beings? Cloning As a Source of Tissue Transplantation. Julian Savulescu. *Journal of Medical Ethics* v. 25 pp87-95 1999.

The subtitle of Savulescu's article states his own position on the subject of human cloning. Savulescu thoroughly examines the most common arguments for and against cloning, including the potential for abuse of the process, its possible role in eugenics, and the manner in which it could infringe upon a person's feeling of individuality. He then explores four hypothetical instances in which human cloning could prove medically beneficial to the seriously ill and concludes that those benefits outweigh the moral and philosophical objections.

Critical Ethical Issues in Clinical Trials with Xenotransplants. Harold Y. Vanderpool. *The Lancet.* v. 351 pp1347-50 May 2, 1998.

Vanderpool argues the case for transplanting animal organs into those awaiting organ transplants. He includes numerous hypothetical cases and examples from both the United States and Europe which consider the recipients' right to privacy, the likelihood of success, and the harm-benefits assessment that must be done for each patient who agrees to undergo this still experimental procedure.

Double Takes. Aline A. Newman. *National Geographic World* pp12-15 Sept. 1999.

Newman discusses the 1996 cloning of Dolly the sheep by Ian Wilmut, a biologist at the Roslin Institute in Edinburgh, Scotland. As he explains, scientists continue to conduct research into cloning, with the hope that it will lead to the development of new medicines, improved food production, and a future for endangered species. The writer discusses the history of cloning and the controversy surrounding the technology. A diagram illustrates how Dolly was cloned.

Genetics of Body-Weight Regulation. Gregory S. Barsh. *Nature* v. 404 pp644-51 April 6, 2000.

Barsh writes that genetics plays a two-fold role in obesity. He explains that many fundamental insights into obesity are provided by studying rare mutations in human beings and model organisms. This research, he says, complements population-based studies that attempt to reveal primary causes. As functional genomics and the human genome project expand and mature, the pace of advances in this area are likely to accelerate. According to Barsh, a possible future convergence of approaches based on Mendelian and quantitative genetics may ultimately lead to more rational and selective therapies.

Impasse. Andy Coghlan. *New Scientist* v. 164 p12 Dec. 11, 1999.

Coghlan reports on the conflicts that arose outside of and within the World Trade Organization (WTO) meeting in Seattle, Washington, in December 1999. While riot police and protestors fought outside the meeting, he explains, inside there was resistance to U.S.-led proposals for the WTO to regulate trade in agricultural biotechnology. According to Coghlan, other European delegations were adamant that the introduction of genetically modified food, plants, and animals should be regulated by the UN's Biosafety Protocol.

Forever Free. John Sulston. *New Scientist* v. 166 pp46-47 April 1, 2000.

A member of the Human Genome Project gives his opinion on why data from the sequencing of the full human genome should be made available to everyone and not be controlled by private companies.

Facts and Friction. Ziauddin Sardar. *New Statesman* v. 128 pp42-44 Nov. 1, 1999.

According to Sardar, people are becoming much more aware of the intricacies regarding scientific issues such as genetically modified (GM) food, and this is mainly due to the Internet. Pressure exerted by an increasingly scientifically literate public has forced American biotechnology corporation Monsanto to back down over the issue of "terminator" seeds, which would produce crops generating only sterile seeds. Sardar argues that it is impossible to know whether gene splicing in plants is safe, and it could well be decades before the true extent of the risks are known. The public's demand that GM foods be proven safe before being released makes good scientific sense, and there are precedents that support this viewpoint. The article asserts that the public should not try to argue scientific technicalities with the industry on the basis of its success with Monsanto, but it should be aware of how politics and all that goes with it is now intrinsically linked with science.

Both Hands and Forearms Transplanted for First Time. Lawrence K. Altman. *New York Times* (Late Edition) pA6 Jan. 15, 2000.

Altman reports on the first double hand-and-forearm transplant, performed in Lyon, France on a 33-year-old Frenchman who lost his hands in a fireworks accident. According to Altman, the international team of surgeons was the same one that performed the first successful hand and forearm transplant, in 1988. Although the patient was reported to be doing well, Altman says, it will be some time before doctors can determine whether the transplant is successful.

The Recycled Generation. Stephen S. Hall. *New York Times Magazine* pp30-35+ Jan. 30, 2000.

According to Hall, stem-cell research has the potential to supply new body parts, but its development is being slowed by abortion politics and corporate rivalries. He explains what stem cells are and their versatility at the embryonic stage, at which time they can become any cell, tissue, or organ in the human body. Theoretically, he says, they could be used to generate an infinite supply of replaceable body parts and could be combined with controversial technologies like cloning to provide a limitless supply of transplant tissue. Hall then discusses the role of right-to-life groups in the debate over the technology and discusses research on stem cells being conducted by Advanced Cell Technology in Worcester, Massachusetts.

Clone of Silence. Margaret Talbot. *New York Times Magazine* pp21-22 April 16, 2000.

Talbot writes about the company Genetic Savings and Clone, which offers to store the DNA of aging dogs and cats so that a similar pet may someday be cloned for the owner, at a cost, initially, of around $200,000. The company is an offshoot of the Missyplicity Project, founded in 1998 when an anonymous couple gave $2.3 million to Texas A & M University to clone the couple's pet dog, Missy. According to Talbot, the Missyplicity Project is a reminder of the way in which the idea of cloning has been normalized, even made cute, in a remarkably short time. She writes that the project represents the first example of cloning for sentimental reasons, as well as the first attempt to re-create a specific animal, reinforcing the belief that loved ones can be replaced by genetic copies.

DNA on Trial. Peter J. Boyer. *The New Yorker* v. 75 pp42-53 Jan. 17, 2000.

Boyer examines the Innocence Project, begun in the early 1990s in lower Manhattan, which aims to use DNA testing to exonerate people who have been wrongly convicted of crimes. According to Boyer, the project has played a role in 39 exonerations. He then reviews the case of Calvin C. Johnson Jr., a black man accused by a white person and convicted by an all-white jury in less than an hour in the South. The writer discusses doubts by Clayton County District Attorney Robert Keller over the innocence of Johnson, who was freed in June 1999 after the Innocence Project carried out DNA testing on the evidence in his case.

Finding a Key to Down Syndrome. Sharon Begley. *Newsweek* p46 May 22, 2000.

Begley addresses how the sequencing of chromosome 21 could help find the cause of Down syndrome. This condition, which can result in mental retardation, heart disease, and the early onset of Alzheimer's, is due to the presence of three copies of chromosome 21 instead of two. As Begley explains, a team of researchers led by labs in Japan and Germany recently put forward the first explanation of why inheriting three copies of any other chromosome causes death before or soon after birth, but three copies of chromosome 21 do not. During their research, she says, the scientists detailed the chemical makeup of the genes that chromosome 21 carries, giving scientists a chance at determining precisely how an extra copy causes Down syndrome.

A Revolution in Medicine. Geoffrey Cowley and Anne Underwood. *Newsweek* pp58-62 April 10, 2000.

Cowley and Underwood examine the kinds of groundbreaking techniques that are being developed with the knowledge gained from DNA tests. One innovation they discuss involves the creation of drugs that will attack the causes of diseases, rather than treat the symptoms. The writers also explain how early detection through genetic testing can determine individual patient response to treatment, thereby enabling doctors to better prescribe a form of treatment. They then write about gene therapy, which they call a "hellishly difficult" technique that is often dangerous to patients. As they explain, increasing the effectiveness of such treatments makes the successful sequencing of the human genome so essential to researchers.

The New Bionic Man. Mike Fillon. *Popular Mechanics* pp50-55 February 1999.

Fillon summarizes the latest developments in artificial body parts and bionic devices. These include not only innovations in artificial muscles and arms, for instance, but also in the means to power them. Among the cutting-edge devices Fillon describes are artificial retinas, cochlear implants, and bionic noses and tongues, as well as devices that prevent epileptic seizures.

New Gene-Therapy Techniques Show Potential. N. Seppa. *Science News* v. 157 p309 May 13, 2000.

According the Seppa, a group researching gene therapy at Stanford University School of Medicine, California, has successfully bypassed the viral approach to gene delivery by using transposons as the gene-delivery vehicles, a technique they hope will avoid a number of serious problems that arise in this form of treatment. Three other research groups, the writer explains, are using another technique to insert large genes into cells, but all of these developments will need extensive animal testing before their true efficacy is confirmed.

Ethics and Embryonic Cells. Roger A. Pedersen. *Scientific American* v. 280 p71 April 1999.

Despite their enormous potential for medical research, Pederson writes, human embryonic stem cells also raise ethical problems because they are obtained from human embryos. As he explains, a congressional ban prohibits the use of federal funds to support any research involving embryos, but in 1994 a panel from the National Institutes of Health recommended that some embryo research was morally justifiable and deserved consideration for federal funding. The writer asserts the legitimacy of embryo research and how its use can avoid any ethical transgressions.

Nanomedicine Nears the Clinic. David Voss. *Technology Review* v. 103 pp60-63+ Jan./Feb. 2000.

Voss looks at nanotechnology, the use of microscopic robots to identify diseased cells in the human body and treat serious illness, a technique that some researchers hope will one day be an alternative to human gene therapy. As Voss explains, doctors like James Baker of the University of Michigan Medical School are working with engineers to develop nanorobots to perform a variety of complex functions. These tasks include identifying and repairing defective genes and functioning as "smart bombs" to "detect pre-malignant and cancerous changes" in living cells, after which they will not only kill those cells but also verify their destruction.

What Will Replace the Tech Economy? Stan Davis. *Time* v. 155 pp76-77 May 22, 2000.

According to Davis, a bioeconomy will replace the present infotech economy a quarter-century from now, as that bioeconomy proceeds through the typical economic life cycle: gestation, growth, maturity, and decline. As he explains, the decoding of the human genome signals the end of the bioeconomy's gestation period, and the growth period will see the emergence of new industries that will combine organic biotech with inorganic silicon infotech, inorganic composite materials, and nanotechnologies. A number of biological processes will be digitized during this stage. Davis asserts that the central growth industries will be pharmaceuticals, health care, agriculture, and food. He claims that beyond the first quarter of the 21st century, biotech will be applied to areas apparently unrelated to biology, and its problems—largely concerning ethics—will be as pervasive as its benefits.

The Sickle-Cell Kid. Frederic Golden. *Time* v. 154 p92 Dec. 20, 1999.

Golden reports on Keone Penn, who, in December 1998, became the first sickle-cell patient to receive a transplant of blood cells from the umbilical cord of a newborn infant. Now aged 13, Penn was born with the most severe form of sickle-cell anemia, a hereditary blood disorder that afflicts over 70,000 Americans, the majority of whom are of African descent. As Golden explains, other young sickle-cell patients have undergone transplants using bone-marrow cells that had to be an exact match with the recipients' own blood. The writer discusses the transplant procedure performed by Dr. Andrew Yeager at Emory University medical school in Atlanta, who recently pronounced Penn cured.

The Last Resort. Jeffrey Kluger. *Time* v. 153 pp68-69 April 26, 1999.

According to Kluger, a controversial breast-cancer therapy has provoked a conflict between women and health insurers. Over the last decade, he writes, more than 12,000 American women have opted for therapy involving the transplant of stem cells that are the precursors of disease-fighting immune-system cells, but many insurers have refused to cover the treatment. Recently, Kluger explains, public opinion, several high-profile lawsuits, and legislation in 10 states have forced insurers to cover transplants. Patients realize, however, that a company can initially agree to cover treatment but later change its mind and that laws requiring reimbursement can also be changed. The debate surrounding the treatment is discussed.

Battle Pending. J. Madeleine Nash. *Time* v. 155 p71 April 17, 2000.

The U.S. Patent and Trademark Office is dealing with a glut of applications for patents on human genes, Nash writes. The race for patents has caused concern that a few powerful corporations will end up controlling invaluable genetic information. According to Nash, trouble has already erupted between university researchers—many of whom are allied with the Human Genome Project—and tenacious firms such as Celera who are rapidly rushing out patent claims. Nash asserts that the private sector needs the money gene patents can generate in order to fuel the biotechnology industry, but many are critical of moves to issue patents on DNA sequences whose commercial use in not yet known.

Will We Still Need to Have Sex? Matt Ridley. *Time* v. 154 pp66-68 Nov. 8, 1999.

Ridley argues that sex will be used more for recreation than procreation in the future. He predicts that many human beings, notably those who are rich, vain, and ambitious, will use test tubes to clone themselves and manipulate their genes, as well as to counteract infertility and the lack of suitable partners. Kluger then refers to the argument by Olivia Judson, author of the book *Dr. Tatiana's Sex Advice for All Creation*, that the public will probably oppose the widespread genetic modification of people as they do genetically modified soya. Even if sex turns out to be genetically unnecessary, Kluger explains, it would not be a total waste of energy: Geoffrey Miller, an evolutionary psychologist at University College London, believes that everything extravagant about human life, from poetry to fast cars, is based on sexual competition.

Will We Clone a Dinosaur? Matt Ridley. *Time* v. 155 pp94-95 April 10, 2000.

Ridley claims that the idea of cloning a dinosaur is not completely farfetched, especially given the numerous species of birds—believed to be descendants of Dinosaurs—that are still in existence. Because, he argues, DNA is turning out to be a great deal more "conserved" than anybody ever imagined and that geneticists are discovering many copies of old, discarded genes in human and animal DNA, the implications are that dinosaur genes are still out there. In a few decades, Ridley writes, some researchers may try to re-create the genome of a dinosaur, from which point it may be possible to devise a simple DNA recipe for a dinosaur creature.

Five Little Piggies Going to Market. Joannie Schrof Fischer. *U.S. News & World Report* v. 128 p51 March 27, 2000.

Fischer reports on five female piglets born in March 2000 in Blacksburg, Virginia, who are the first cloned members of their species. She says that the piglets offer the most positive hope yet of saving the lives of the 63,000 people in the United States who are awaiting organ transplants. Fischer then explains what makes pigs good organ donors for human beings, as well as the potential susceptibility to serious disorders like AIDS which may accompany such transplants. According to Fischer, Robert Lanza, coauthor of *Xeno: The Promise of Transplanting Animal Organs Into Humans*, says that the DNA in a cloned pig embryo could be manipulated to make the organs more humanlike.

How to Build a Better Bull. Leslie Roberts. *U.S. News & World Report* v. 128 p51 Jan. 17, 2000.

Roberts writes that scientists at the University of Connecticut have developed a new way to clone animals from cells that have been grown in a laboratory. Cattle cloning is becoming more widespread, she says, but the new method would allow scientists to experiment with the cells' genes before creating a new cow. The process involves scraping skin cells from the ear of a bull and growing the cells in a culture. As Roberts explains, however, the researchers have warned that exact manipulation of bovine genes has not yet been attempted.

Showing Some Spine. Stacey Schultz. *U.S. News & World Report* v. 127 p67 Dec. 13, 1999.

Schultz discusses the development of a therapy that shows promise for people who have been paralyzed as a result of spinal-cord injuries. Scientists from Washington University School of Medicine in St. Louis report in the December *Nature Medicine* that they have transplanted mouse embryonic stem cells into rats' injured spines. As Schultz explains, these undifferentiated cells helped repair the animals' injured spinal cords by replacing the missing nerve cells, allowing the paralyzed rats to walk again.

Of Genes, Grain, and Grocers. Laura Tangley. *U.S. News & World Report* v. 128 pp49-50 April 10, 2000.

Tangley writes that it is unclear whether consumers' fears regarding the safety of genetically engineered crops is justified. She claims that, for the first time since genetically modified seeds were first used, in 1996, numerous American farmers are likely to reduce the amount of land they plant with genetically modified crops because consumers, who are worried that the crops could damage human health and the environment, might refuse to purchase them. Tangley claims that early results indicate that a number of fears about engineered crops are overstated: There is no proof that any genetically modified food has adversely affected a human being. Nonetheless, she acknowledges, the results also suggest that other concerns, especially about the environmental impact of these crops, might be well-founded; for instance, genes that are meant to provide crops with a competitive advantage might be passed to related wild plants with which they interbreed, resulting in new "superweeds."

DNA in the Dock. Martine Jacot. *UNESCO Courier* v. 53 pp37-9 April 2000.

According to Jacot, although genetic fingerprinting offers almost flawless proof and is helping identify criminals and free people wrongly convicted of crimes, it could violate fundamental human rights. Jacot discusses British scientist Alex Jeffreys' work with DNA identification but explains that identifying a person's "traces" at a scene of a crime does not automatically mean that he or she is guilty. The writer reviews the creation of a DNA database by British and American police and explores the question of whether or not a DNA sample can be taken from someone without permission. Jacot writes that police can analyze DNA from a confiscated toothbrush without a person's knowledge and, they claim, without violating recognized concepts of human rights. She concludes that the easier it becomes to establish genetic profiles from DNA, the more urgent is the need to safeguard against the possible misuse of the genetic identifications and samples that are preserved.

Dolly's Premature Aging Not Found in Cloned Cows. Rick Weiss. *Washington Post* pA3 April 28, 2000.

In his article, Weiss discusses one byproduct of the cloning of cows that had not been anticipated: a lack of premature aging found previously in Dolly, the cloned sheep. This discovery, he writes, provides an intriguing prospect for individuals searching for an antidote to old age. Weiss reports on speculations by researchers that, through a procedure called therapeutic cloning, the cells from a cloned embryo could be used to replace older, dying cells and essentially restore a person's lost youth. As Weiss relates, news of this discovery and its proposed applications have met with considerable resistance in the United States, where scientists continue to study the cloning process in the hopes of achieving a similar result by less controversial means.

Index